Enikö Kokai

C-Myc expression in adult and embryonic endothelial cells

AF062745

Enikö Kokai

C-Myc expression in adult and embryonic endothelial cells

C-Myc is essencial for adequate vasculogenesis and angiogenesis

Südwestdeutscher Verlag für Hochschulschriften

Imprint
Any brand names and product names mentioned in this book are subject to trademark, brand or patent protection and are trademarks or registered trademarks of their respective holders. The use of brand names, product names, common names, trade names, product descriptions etc. even without a particular marking in this work is in no way to be construed to mean that such names may be regarded as unrestricted in respect of trademark and brand protection legislation and could thus be used by anyone.

Publisher:
Südwestdeutscher Verlag für Hochschulschriften
is a trademark of
Dodo Books Indian Ocean Ltd., member of the OmniScriptum S.R.L Publishing group
str. A.Russo 15, of. 61, Chisinau-2068, Republic of Moldova Europe
Printed at: see last page
ISBN: 978-3-8381-2425-4

Zugl. / Approved by: Ulm, Ulm University, Diss., 2010

Copyright © Enikö Kokai
Copyright © 2011 Dodo Books Indian Ocean Ltd., member of the OmniScriptum S.R.L Publishing group

CONTENTS

SUMMARY 5
1. INTRODUCTION 7
   1.1. Structure and function of the vascular system 7
   1.2. Development of the vascular system 11
      1.2.1 Vasculogenesis 11
      1.2.2 Angiogenesis 12
      1.2.3 Pruning and remodeling 13
      1.2.4 Maturation and remodeling 13
      1.2.5 Lymphangiogenesis 14
   1.3. Molecular regulation of vascular development 16
      1.3.1 Establishing the embryonic vasculature 16
      1.3.2 Molecules regulating blood vessels branching, remodeling, maturation ... 18
   1.4. The C-Myc proto-oncogene 21
      1.4.1 Structure, function and transcriptional regulation of c-Myc 22
      1.4.2 C-Myc and cell proliferation 24
      1.4.3 Gene targets of c-Myc 24
   1.5. C-Myc in vascular development 26
THE AIMS OF THE STUDY 29
2. MATERIALS AND METHODS 31
   2.1. Transgenic mice 31
   2.2. Animal experiments 32
      2.2.1 Gross embryonic pathology 32
      2.2.2 Gross pathology of adult animals 32
      2.2.3 Doxycycline treatment of tumor bearing adult mice 32
      2.2.4 Tumor cell line transplantation 32
   2.3. Genotyping of transgenic mice 33
      2.3.1 Genomic DNA isolation 33
      2.3.2 Polymerase Chain Reaction (PCR) 33
   2.4. Detection of gene expression 34
      2.4.1 Reporter gene detection 34
      2.4.2 Analysis of mRNA expression levels 34
      2.4.3 Quantitative real-time PCR analysis 36
   2.5. Protein analysis 36
      2.5.1 Protein concentration measurement by Bradford method 36

- 2.5.2 Western immunoblot ... 37
- 2.5.3 VEGF-A ELISA ... 37
- 2.5.4 Total MMP-9 ELISA ... 37
- 2.6. Flow cytometry analysis ... 37
  - 2.6.1 Isolation of primary embryonic endothelial cells ... 38
  - 2.6.2 Detection of apoptosis and proliferation by flow cytometry ... 38
  - 2.6.3 Fluorescence Activated Cell Sorting (FACS) ... 38
- 2.7. Histological analyses ... 39
  - 2.7.1 Immunohistochemistry on tissue sections ... 39
  - 2.7.2 Immunohistochemistry on tissue peaces: Whole-mount analyses ... 40
  - 2.7.3 Electron Microscopy ... 42
- 2.8. Quantitative analysis of immunohistological stainings ... 43
  - 2.8.1 Staining quantification ... 43
  - 2.8.2 Quantitative analysis of dermal vascular architecture ... 43
- 2.9. Materials ... 45
  - 2.9.1 General chemicals ... 45
  - 2.9.2 General buffers and solutions ... 45
  - 2.9.3 Histology chemicals and materials ... 50
  - 2.9.4 Special consumption items and equipment ... 51
  - 2.9.5 Antibodies ... 53
  - 2.9.6 The list of primers used for RT-PCRs ... 54
- 3. RESULTS ... 57
  - 3.1. Model for conditional c-Myc expression in endothelial cells ... 57
  - 3.2. Endothelial cell-specific c-Myc expression in adult mice ... 59
    - 3.2.1 Characterisation of the Tie2-tTA/tetO-Myc adult animals ... 59
    - 3.2.2 Transgene expression in Tie2-tTA/tetO-Myc adult animals ... 59
    - 3.2.3 Survival of Tie2-tTA/tetO-Myc adult animals ... 59
    - 3.2.4 Gross pathology of Tie2-tTA/tetO-Myc animals ... 61
    - 3.2.5 Tumor classification and statistics ... 62
    - 3.2.6 Tumor regression upon transgene inactivation ... 66
    - 3.2.7 Establishment of tumor cell lines ... 66
    - 3.2.8 Characterization of the tumor cell line ... 67
    - 3.2.9 Subsummary I ... 70
  - 3.3. Characterisation of Tie2-tTA/tetO-Myc embryos ... 72
    - 3.3.1 Transgene expression in Tie2-tTA/tetO-Myc embryos ... 72

- 3.3.2 Endothelial cell-specific c-Myc expression causes embryonic lethality 74
- 3.3.3 Possibly origins of vascular permeability 77
- 3.3.4 Analysis of the lymphatic vessels 80
- 3.3.5 Analyses of dermal blood vessel architecture 83
- 3.3.6 Electron microscopical analysis of embryonic endothelium 87
- 3.3.7 Quantification of endothelial cell apoptosis and proliferation 89
- 3.3.8 Isolation of embryonic endothelial cells 91
- 3.3.9 Angiogenic modulators expression in purified embryonic endothelial cells 94
- 3.3.10 Subsummary II 97

4. DISCUSSION 99
4.1. C-Myc expression in endothelial cells of adult mice 99
- 4.1.1 C-Myc expression in Tie2-manner in adult mice 99
- 4.1.2 C-Myc inactivation *in vivo* 99
- 4.1.3 Mouse model for human diseases 100
- 4.1.4 Animal model for Kaposi's sarcoma 101

4.2. C-Myc expression in embryonic endothelial cells 101
- 4.2.1 Possible origins of vascular permeability 101
- 4.2.2 Vascular permeability protecting factors 105
- 4.2.3 Factors modulating vessel remodeling 106
- 4.2.4 C-Myc and angiogenic switch 107
- 4.2.5 C-Myc: apoptosis and/or proliferation? 108
- 4.2.6 The role of c-Myc in embryonic vascular development 109
- 4.2.7 Proposed model of the role of c-Myc during embryonic vascular development 111

ABBREVIATIONS 113
REFERENCES 115

SUMMARY

Previous work had shown that the transcription factor c-Myc is required for normal vasculogenesis and angiogenesis during embryonic development. To further investigate the contribution of c-Myc to these processes, we conditionally over-expressed c-Myc in adult and embryonic endothelial cells using the tetracycline-regulatable system, the so-called tet-off system, blocking the transgene expression in the presence of doxycycline.

Vascular endothelial cell-specific overexpression of c-Myc in adult mice induced pathological malformations resulting in angiosarcomas and/or adenomas. Starting from week 22 adult double transgenic mice developed tumors and died in average with 36 weeks. Inactivation of the transgene system by doxycycline *in vivo* results in partial tumor regression.

We established one angiosarcoma cell line, expressing these endothelial and mesenchymal markers. However, during cell culture propagation the cell line lost its regulatable feature by doxycycline. Under *in vivo* conditions the tumor cell line was no longer regulatable by doxycycline application, when it was subcutainiously transplantated in Rag2-\- mice developed xenografts independent of doxycycline administration.

The endothelial c-Myc overexpression during embryogenesis resulted in severe defects in the vascular system. The c-Myc expressing embryos died between embryonic day (E) E14.5 and E17.5 and suffered from widespread edema formation and multiple hemorrhagic lesions. The changes in vascular permeability were not caused by deficiencies in vascular basement membrane composition or pericyte coverage. However, the overall turnover of endothelial cells was elevated and revealed by increased levels of proliferation and apoptosis. With whole mount immunohistochemical analysis we revealed alterations in the architecture of capillary networks. The dermal vasculature of c-Myc expressing embryos was characterized by a reduction in vessel branching, which occured despite the up-regulation of the pro-angiogenic factors VEGF-A and angiopoietin-2 (Ang-2). Thus, the net outcome of an excess of VEGF-A and Ang-2 in face of an elevated cellular turnover appears to be a defect in vascular integrity.

1. INTRODUCTION

1.1. Structure and function of the vascular system

The vascular system is composed of a multitude of branched vessels that carry blood and lymph through the body.

Blood vessels supply the body with O_2 and CO_2 as well as with nutrients, and it transports the metabolic waste products (Schmidt 2000). Other essential functions of the blood vascular system are the carriage of hormones and the molecules of the immune defense, furthermore the osmoregulation and the thermoregulation of the body.

The lymphatic vessel system functions parallel to the blood vascular system (Schmidt 2000). Its function is the collection and back transport of proteins and other molecules from the interstitium into the blood. The lymph flows unidirectional.

The blood vascular system includes different blood vessels like arteries, veins, and capillaries. The heart muscle pumps oxygenated blood via arteries to the capillaries where bidirectional exchange of gases and metabolites occurs between blood and tissues. Veins collect deoxygenated blood from the microvasculature and convey it back to the heart (Alberts 2001).

The composition of the different vessel types is adequate to their function. Arteries have a thick wall of connective tissue and many layers of vascular smooth muscle cells (vSMC or SMC) (Figure 1) (Hellstrom 1999). In contrast, the veins have few smooth muscle cells in the wall to conduct pressure. They are thin-walled vessels, with large-diameter and many valves for preventing blood backflow. The finest branches of the vascular tree are the capillaries. Capillaries are thin-walled vessels, where the wall consists of only one vascular endothelial cell layer, covered with a basement membrane and a few scattered pericytes (PC). Pericytes are cells of the connective-tissue family, related to vSMCs that wrap themselves round the small vessels and supply mechanical stability (Gerhardt 2003).

Common feature of all vessels types is the inner layer of a single sheet of vascular endothelial cells, the endothelium. The endothelium functions as a selective barrier between the vessel lumen and surrounding tissue, controlling the diffusion of gases and metabolites into and out of the bloodstream. The endothelial cell layer is separated from the underlying connective tissue through a specialized extracellular matrix (ECM), the basement membrane (Partridge 1992).

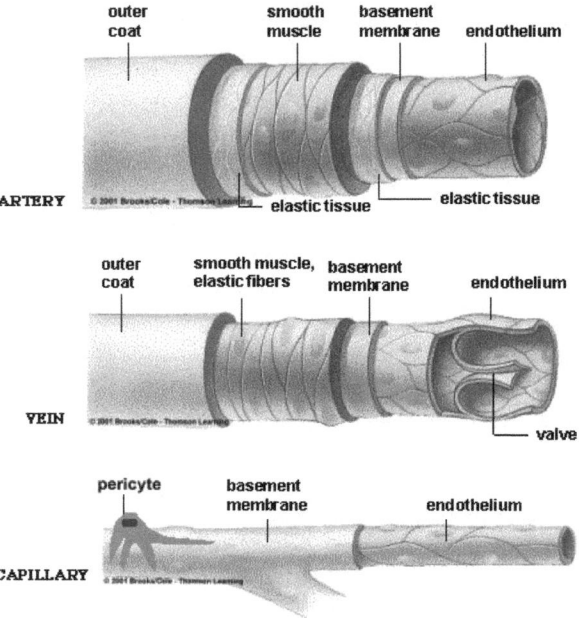

Figure 1. The structue of blood vessels.
Arteries have a thick, muscular wall to carry blood from the heart to the body. Veins are thin-walled vessels with many valves to prevent backflow. Capillaries are thin-walled and very small in diameter. Common to all vessel types are the endothelial cell layer and the basement membrane. Vessels are surrounded by vSMCs and pericytes, which form one or multiple layers increasing in thickness with vessel size. Figure modified from (Solomon 2005).

Endothelial cells adhere to each other through junctional structures formed by transmembrane proteins that are responsible for homophilic cell-to-cell adhesion. The transmembrane proteins are linked to specific intracellular partners, which mediate anchorage to the actin cytoskeleton and, as a consequence, stabilize junctions. There are two types of endothelial cell-to-cell adhesion junctions called adherens junctions (AJ) and tight junctions (TJ) (Figure 2) (Saitou 2000; Dejana 2008). The primary function of these junctions in endothelial cells is to maintain tissue fluid homeostasis.

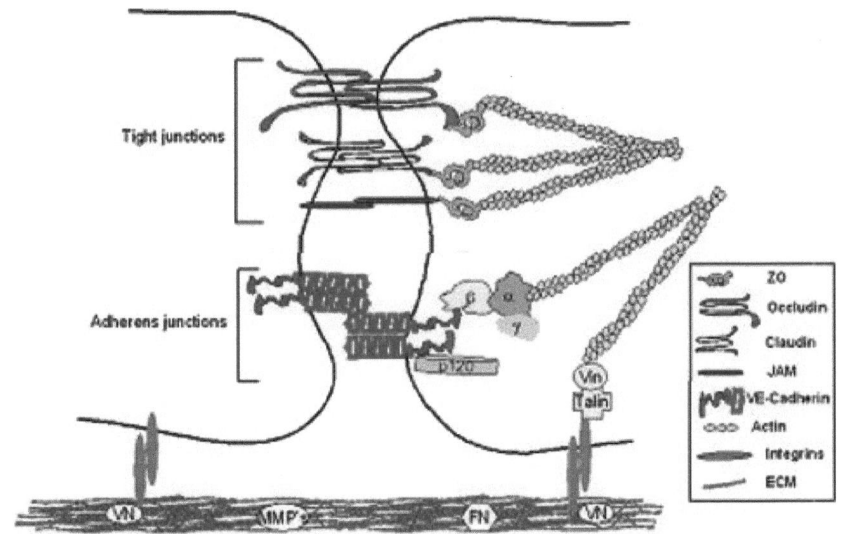

Figure 2. Structural organization of endothelial cell junction.
This figure shows a typical arrangement of the different endothelial cell junctions and integrin receptors by which endothelial cells adhere to each other and to ECM, respectively. Occludin, claudins, and JAMs are the backbones of TJs, whereas VE-cadherin is required for formation of AJs. While the extracellular domains of occludin, claudins, and VE-cadherin maintain cell-cell contact, intracellular domains provide junctional stability through their linkages with the actin cytoskeleton via catenins (α-, β-, γ-catenin, p120-catenin) or zonula occludens-1 protein (ZO-1). Integrin receptors link endothelial cells with ECM through matrix proteins which are fibronectin (FN) or vitronectin (VN). The cytosolic domains of integrins are linked with the actin cytoskeleton through the proteins talin and vinculin (Vin) involved in integrin-mediated signaling. MMPs (membrane metalloproteases) control remodeling of ECM and modify cell-ECM signaling. Figure modified from (Mehta 2006).

The endothelial cells create a barrier, thus controlling the passage of plasma proteins and circulating cells from the blood to the tissue. This function is achieved through the so-called transcellular and paracellular pathways, which means solutes and cells can pass through (transcellular) or between (paracellular) endothelial cells (Bazzoni 2004). Transcellular pathway comprises the passage of plasma components through the endothelial cytoplasm by the action of vesicular systems and fenestrae (Roberts 1995; Engelhardt 2004; Millan 2006a; Millan 2006b; Nieminen 2006). By contrast, the paracellular pathway is mediated by the coordinated opening of cell-to-cell junctions (TJs) and/or by rearrangement of their architecture. This function must be tightly regulated to maintain endothelial integrity and to prevent exposure of the subendothelial matrix of blood vessels (Vestweber 2007). Although both TJs and AJs can play an important role in the control of endothelial permeability, TJs have always been considered as the key regulators of this function (Bazzoni 2004).

AJs are important for vascular development and for correct organization of new vessels in angiogenesis. One of the main components of endothelial AJs is vascular endothelial VE-cadherin. VE-cadherin mediates homophilic adhesion between adjacent endothelial cells. Catenins are components in the AJs and they associates with VE-cadherin through its cytoplasmic domain (Nyqvist 2008). Deletion of VE-cadherin in mice is embryonally lethal (E9.5) due to immature vascular development

The lymphatic vascular system is composed of lymphatic vessels, lymph nodes, and lymphocytes, is a distinctive vasculature, different yet similar to the blood vasculature (Witte 2006). The lymphatic vessels drain protein-rich lymph and immune cells from tissue spaces and return them to the blood circulation (Witte 2006).

The inner site of lymphatic vessels is lined with a specific type of endothelial cells, called lymphatic endothelial cells. Lymphatic capillaries lack a continuous basement membrane, vascular mural cells and tight junctions therefore they are highly permeable (Figure 3). Confocal microscopic imaging studies demostrated discontinuous junctions in the lymphatic capillary endothelium where fluid can enter without repetitive formation and dissolution of intercellular junctions (Baluk 2007). Elastic fibers known as anchoring filaments connect lymphatic capillary endothelial cells to the surrounding stroma (Oliver 2004). Lymph passes through the capillaries to the collecting lymphatic vessels and finally to the thoracic duct. The larger lymphatic vessels contain connective tissue and contractile smooth muscle cells (Figure 3). They have valves to prevent backflow. The valve regions are devoid of vSMCs. The thoracic duct drains lymph back into the blood circulation at the left subclavian vein (Oliver 2005).

Lymphatics are involved in diverse developmental, growth, repair, and pathological processes (Witte 2006). Pathological disorders like, edema can arise due to dysfunction of the lymphatic vasculature and insufficient lymph drainage. It results in the accumulation of protein rich fluid in tissues causing swelling of the extremities, a symptom called lymphedema (Witte 2006). Another form of lymphatic dysfunction occurs when lymph extravasates and accumulates in the abdomen. This disorder is called chylous ascites and is a result of abnormal lymphatic vessel development (Aalami 2000; Jeon 2008).

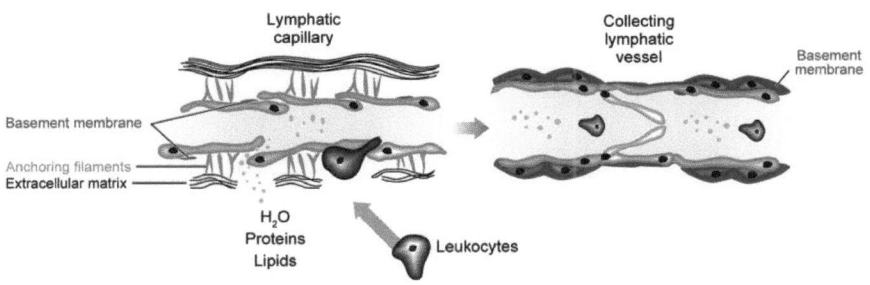

Figure 3. Structure of the lymphatic vessels.
The endothelial cells of lymphatic capillaries (green) lack tight junctions. Instead, the neighboring endothelial cells partly overlap, forming valve-like openings, which allow easy access for fluid, macromolecules, and cells into the vessel lumen. The lymph drains from the lymphatic capillaries to collecting lymphatic vessels, which are emptied into veins in the jugular region. Collecting lymphatic vessels have a basement membrane, are surrounded by vSMCs (red) with intrinsic contractile activity to promote lymph flow. They contain valves that prevent backflow of the lymph. Figure modified from (Karpanen 2008).

1.2. Development of the vascular system

1.2.1 Vasculogenesis

The cardiovascular system is the first functional organ system that develops in the embryo. The embryonic survival, growth and differentiation are depending on transport of nutrients and waste products through the early vasculature.

During embryonic development the first blood vessels evolve by vasculogenesis (Risau 1995). During vasculogenesis vascular endothelial cells differentiate from their precursors, called angioblasts, following *de novo* primitive blood vessel formation in extra-embryonic tissues and within the embryo (Risau 1997).

The initial step of vasculogenesis occurs when the common mesodermal precursor cells of both endothelial and blood cells, the hemangioblasts, form aggregates called blood islands (Figure 4) (Risau 1997). Cells located in the middle of a blood island develop into hematopoietic precursors around E7 (Rossant 2002). Thereby the outer cell population develops into endothelial progenitor cells, called angioblasts (Risau 1997). The first primitive blood vessels evolve *de novo* from these angioblasts and form a primary capillary plexus around E7.5 (Rossant 2002).

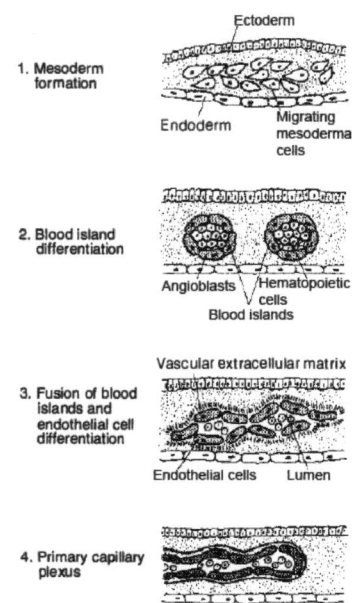

Figure 4. Schematic representation of the processes occurring during vasculogenesis.
Mesoderm formation, blood island differentiation, fusion of blood islands and endothelial cell differentiation, primary capillary plexus formation. Adopted from (Risau 1995).

Studies in avian and quail embryos have shown that basic fibroblast growth factor (bFGF) mediates the induction of angioblasts from mesoderm (Poole 2001). Key molecular regulators of early vascular development are vascular endothelial growth factor VEGF-A, and its cognate receptors, VEGFR-1 and VEGFR-2. It was found that the earliest marker defining angioblasts is the expression of VEGFR-2 at E7 (Kappel 1999). It is known that VEGF-A influences angioblast differentiation through its receptor VEGFR-2 (Shalaby 1995; Carmeliet 1996; Ferrara 1996). In addition, it was found that the lack of VEGFR-1 suppresses the hemangioblast commitment in VEGFR-1 knock out mice (Fong 1999).

1.2.2 Angiogenesis

After the uniform primary vascular plexus has been formed, new capillary structures are generated from the pre-existing vessels in a process termed angiogenesis (Risau 1997). Angiogenesis takes place from E8 onward (Rossant 2002).

Angiogenesis is defined as the formation of new vascular structures from preexisting vessels by sprouting and intussusception (Partanen 1999). During sprouting, proteolytic degradation of the extracelllar matrix is followed by chemotactic migration and proliferation of endothelial cells, formation of a lumen and functional maturation of the endothelium. VEGF-A is a well known angiogenesis-activating factor, inducing one or more of these activities in endothelial cells (Risau 1997). Intussusception happens when endothelial cells sprout into the lumen of the vessel and produce transcapillary pillars that traverse the lumen. These pillars fuse with the opposite side of the vessel, thereby creating a vascular loop (Partanen 1999).

Mouse knock-out studies indicate that the Tie1 and Tie2 receptors with their ligands Angiopoietin-1 and -2 play a role in angiogenic expansion and survival of the endothelium (Dumont 1994; Sato 1995; Suri 1996; Maisonpierre 1997; Puri 1999).

1.2.3 Pruning and remodeling

The emerging vascular plexus adjusts to changes in blood flow and oxygen demand of the tissue by remodeling (Gerhardt 2003). Remodeling involves both new vessel growths by sprouting or intussusception, as well as vessel regression, a process called pruning. These processes result in a mature, hierarchically organized vascular pattern with larger and smaller vessels that facilitates directional blood flow (Risau 1997; Gerhardt 2003).

Molecules, directing the process of vascular remodeling are Ang-1, Ang-2 and VEGF-A.

Ang-1 acts to stabilize vessels. Ang-2 is expressed at sites of vascular remodeling and acts to locally block Ang-1 action and destabilize vessels (Maisonpierre 1997). If VEGF-A is around, then local angiogenesis can occur. In the absence of VEGF-A and in the presence of Ang-2 vessels destabilize and endothelial cells apoptose, resulting in vessel regression (Risau 1997; Rossant 2002).

1.2.4 Maturation and remodeling

The further maturation of blood vessels involves processes, which induce either maturation or regression of vessels along with the tissue or organ they supply (Risau 1997).

The direction of blood flow influences the decision of a blood vessel becoming venules or arterioles. Non-perfused blood vessels regress. Shear stress also strongly affects endothelial cells, inducing modification of cell-cell as well as cell-extracellular-matrix junctions (Risau 1997).

Associated support cells become recruited to the developing vessels, such as smooth muscle cells and pericytes. This phase of vascular development involves the formation of vascular basement membranes too (Carmeliet 2000). Vascular mural cells are very important for vessel maturation and stabilization. Recruitment of mural cells leads to a quiescent state of the endothelium, decreases permeability, gives structural support and provides elasticity for the vessel wall (Armulik 2005).

Platelet-derived growth factor (PDGF-β) signaling was found to be critical for the proliferation and migration of smooth muscle cells and pericytes to the developing vessel wall (Lindahl 1998).

1.2.5 Lymphangiogenesis

The development of lymphatic vessels proceeds parallel, but secondary to the development of the blood vascular system (Figure 5) (Oliver 2004). Blood venous endothelial cells acquire a lymphatic endothelial cell phenotype by the stepwise expression of different gene products. The development of the mammalian lymphatic vasculature takes place in four stages (Oliver 2004).

1) Lymphatic endothelial cell competence describes the process when venous endothelial cells become competent to respond to a lymphatic-inducing signal. Beyond E9.5 all venous endothelial cells in the anterior cardinal vein express lymphatic vessel endothelial hyaluronan receptor 1 (LYVE-1).

2) However, a restricted subpopulation of endothelial cells on one side of the vein around E10.5 starts to express the lymphatic endothelial cell (LEC) specific transcription factor Prox1. These Prox1-expressing LEC progenitors start to bud from the vein in a polarized manner. This process is called lymphatic endothelial bias.

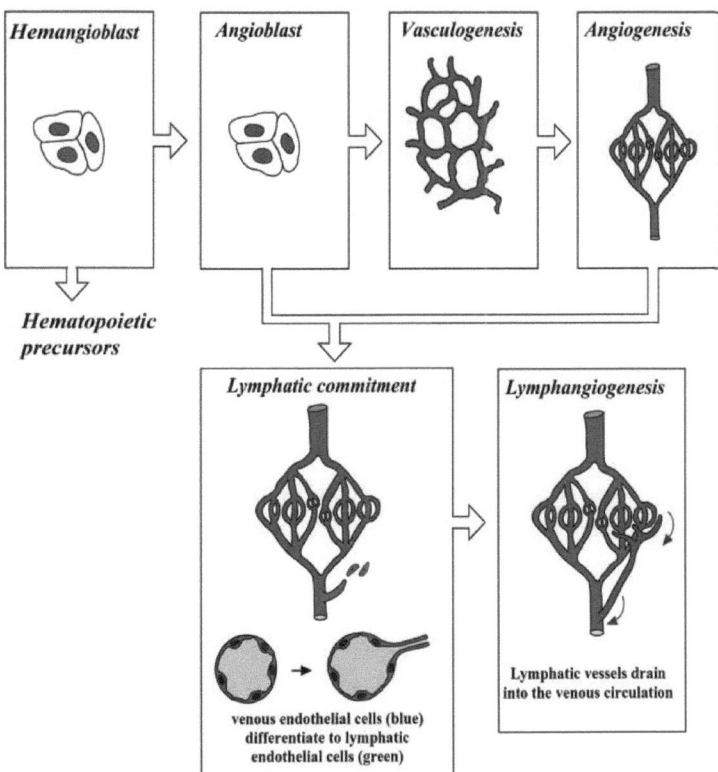

Figure 5. Schematic illustration summarizing the development of the blood and lymphatic vascular systems

Upper panels: The common hematopoietic and endothelial cell precursors, the hemangioblasts, differentiate into endothelial precursors (angioblasts). Prespecified arterial (red) and venous (blue) endothelial precursors proliferate, migrate and form a primitive capillary network (vasculogenesis), which subsequently remodels and expands to a mature hierarchically organized vascular network of arteries and veins (angiogenesis). Lower panels: during embryonic development, venous endothelial cells respond to putative lymphangiogenic signals and differentiate into lymphatic endothelial cells (green, lymphatic commitment). Lymphatic vessels further sprout, expand, and remodel and establish an open-ended vessel system that connects to the venous circulation (lymphangiogenesis). Figure modified from (Alitalo 2002).

3) Specification of lymphatic endothelial cells means that LEC progenitors become mature LECs with the desired LEC phenotype. The simultaneous expression of LYVE-1, Prox-1, VEGFR-3 and secondary lymphoid chemokine (SLC) indicates irreversible commitment to the lymphatic endothelial cell lineage.

4) Subsequently these budding LECs proliferate and migrate to form the embryonic lymph sacs and the lymphatic vascular network. This process is called LEC differentiation. VEGFR-3 expression is maintained at a high level in the budding lymphatic endothelial cells, while its expression becomes weaker in the blood vasculature.

1.3. Molecular regulation of vascular development

1.3.1 Establishing the embryonic vasculature

The members of the Vascular Endothelial Growth Factor (VEGF) family play critical roles in the growth of vascular as well as lymphatic endothelial cells (Rossant 2002). Although the VEGF family includes 5 members, only two members appear to be essential: VEGF-A for vasculogenesis and angiogenesis and VEGF-C for embryonic lymphangiogenesis.

Vascular endothelial growth factor receptors bind VEGFs and mediate their signals (Rossant 2002). There are three high affinity cell surface receptor tyrosine kinases known, called VEGFR-1, VEGFR-2, and VEGFR-3.

VEGF-A

Vascular endothelial growth factor A (VEGF-A) also known as Vascular Permeability Factor (VPF) is one of the most important regulators of endothelial cell physiology (Senger 1983). It mediates developmental, physiological and pathological angiogenesis and induces vascular permeability. During development, it stimulates vascular endothelial cell proliferation and migration and is required for the survival of endothelial cells in newly formed blood vessels (Connolly 1989; Alon 1995; Benjamin 1999). Heterozygous deletion of VEGF-A in embryos shows a lethal phenotype due to abnormal formation of blood islands and blood vessels (Carmeliet 1996; Ferrara 1996). Embryos lacking both *Vegf-a* alleles show a more severe phenotype and die earlier, suggesting that fine-tuned regulation of VEGF-A is essential for a correct early vascular differentiation (Shalaby 1995; Fong 1999).

In addition to its strong angiogenic effect, VEGF-A overexpression induces enlargement of lymphatic vessels in mouse skin (Nagy 2002). During wound healing and in some experimental tumors, VEGF-A induces lymphangiogenesis (Hong 2004; Hirakawa 2005). Further, these studies suggest that VEGFR-2, VEGF-A's main receptor, is expressed in some mainly collecting lymphatic vessels (Hong 2004; Hirakawa 2005). However, with the current knowledge about VEGF-A's role in lymphangiogenesis, it is thought that at least some of the lymphatic effects of VEGF-A might be secondary, due to the induction of vessel permeability and recruitment of inflammatory cells that produce lymphangiogenic growth factors (Cursiefen 2004; Baluk 2005).

Vegf gene expression is regulated by a variety of stimuli such as growth factors like epidermal growth factor, transforming growth factor-β1 (TGF-β1), tumor promoters like *c-myc* and hypoxia (Takahashi 2005). Hypoxia-induced transcription is mediated by the transcriptional activator hypoxia-inducible factor-1 (HIF-1).

VEGF-C

Mouse models suggest that VEGF-C is required for lymphangiogenesis (Kaipainen 1995). In the developing mouse embryo, VEGF-C and VEGFR-3 show adjacent expression patterns in regions where lymphatic vessels develop by sprouting from the embryonic vein (Kaipainen 1995). VEGF-C is essential for the initial sprouting and directed migration as well as for the subsequent survival of lymphatic endothelial cells (Karkkainen 2004). *Vegf-c* knock out mice show embryonic lethality between E15.5-E17.5, and mice with a heterozygous deletion of *vegf-c* display defects in lymphatic vascular development (Karkkainen 2004). Hence, a tight regulation of VEGF-C expression is needed during embryonic development. VEGF-C has been shown to specifically induce growth of lymphatic vessels; however it does not induce blood vessel growth in mouse skin (Jeltsch 1997).

VEGFR-1 (or Flt1) is already expressed in hemangioblasts during embryonic development. Later, VEGFR-1 is expressed also on vSMCs and monocytes/macrophages, besides blood vascular endothelial cells (Hiratsuka 1998; Shibuya 2006). VEGFR-1 functions in monocytes/macrophages as a receptor mediating the migration of these cells (Barleon 1996; Clauss 1996; Hiratsuka 1998; Sawano 2001; Shibuya 2006). Indeed VEGFR-1 shows higher affinity to VEGF-A than VEGFR-2 (Rossant 2002). VEGFR-1 overexpression in endothelial cells does not induce cell proliferation in the presence of VEGF-A (Seetharam 1995; Petrova 1999). *Vegfr1* deficient mice die early in development at E8.5 and show disorganized and malformed primary blood vessels (Fong 1999). The lethal phenotype is achieved through an overgrowth of endothelial cells and increased hemangioblast commitment, which suggests that VEGFR-1 functions as a negative regulator of VEGF\VEGFR signaling during embryonic development (Fong 1995; Fong 1999).

VEGFR-2 (or KDR/Flk1) is already expressed on mesodermal cells at E7.0 and becomes restricted to blood island hemangioblasts, precursors of both endothelial and hematopoietic cells (Shalaby 1995). In adult mice, VEGFR-2 is expressed in endothelial cells and in hematopoietic stem cells as well (Ziegler 1999).

Embryos lacking the *Vegfr2* die around E9 and show no development of any blood vessels or hematopoietic cells indicating VEGFR-2 essential role in the differentiation of endothelial progenitors (Shalaby 1995). Through VEGF-A binding VEGFR-2 is the major signal transducer during angiogenesis, mediating proliferation, migration and survival signals in endothelial cells (Shibuya 2006). In addition, VEGFR-2 is expressed in endothelial tip cells, which sense VEGF-A concentration gradients through their filopodia and guide vascular sprouting (Gerhardt 2003).

VEGFR-3 (or Flt4) is expressed during the early stages of embryonic development in endothelial cells. Later, its expression decreases in blood vascular endothelial cells and becomes restricted to lymphatic endothelial cells (Kaipainen 1995). VEGFR-3 is not required for vasculogenesis, however genetic disruption of VEGFR-3 results in defective remodeling of the primary vasculature, cardiovascular failure, disturbed hematopoiesis and embryonic death by E9.5 (Dumont 1998). VEGFR-3 not only mediates the sprouting and migration of differentiated lymphatic endothelial cells from the veins but is required for the survival and maintenance of lymphatic vessels as well (Makinen 2001a; Karkkainen 2004). VEGFR-3's substrates are VEGF-C and VEGF-D (Rossant 2002).

1.3.2 Molecules regulating blood vessels branching, remodeling, maturation

The homogenous capillary plexus has to undergo extensive remodeling and finally form a highly branched hierarchical vascular network. The establishment of the branching pattern of the vascular system requires coordinated guidance cues from many different molecules (Rossant 2002). The initial formation of vascular networks proceeds independently of perivascular cells. However, subsequent vessel remodeling and maturation relies on mesenchymal cell - endothelial cell signaling (Lindahl 1998).

Ang/Tie family

The Angiopoietin/Tie system acts as a vascular specific ligand/receptor system to control endothelial cell survival and vascular maturation (Augustin 2009).

Angiopoietins

Four Angiopietins are known. The best characterized Angiopoietins are Angiopoietin-1 (Ang-1) and Angiopoietin-2 (Ang-2). Ang-3 and Ang-4 are orthologs found in mouse and human, respectively (Rossant 2002; Augustin 2009). Ang-1 and Ang-2 are ligands for Tie2.

Ang-1 is primarily expressed by mesenhymal cells and acts in a paracrine manner on the endothelium (Augustin 2009). It is abundantly expressed by the myocardium during early development and by perivascular cells later during development and in the adult tissues. Ang-1 is also expressed by tumor cells and neuronal cells of the brain (Augustin 2009).

Ang-2 is almost exclusively expressed by endothelial cells where it is stored in Weibel-Palade bodies (WPB) (Augustin 2009). Upon cytokine activation of the endothelium, Ang-2 is rapidly released from WPB. It acts in an autocrine manner on the Tie2 receptor. Recent studies have shown that endogenous Ang-2 may act through an internal autocrine loop mechanism (Scharpfenecker 2005).

Under physiological conditions, Ang-2 is expressed in regions of vascular remodeling, e.g. during vasularization of the retina (Augustin 2009). Ang-2 expression is upregulated under pathological conditions, e.g., in the endothelium of tumors, and in tumor cells. Ang-2 levels are upregulated by hypoxia. Moreover, retinal neurons and Müller cells are source of Ang-2.

Tie1 and Tie2 receptors

Tie1 is almost exclusively expressed by endothelial cells (Augustin 2009). During embryonic development *tie1* becomes transcriptionally activated in differentiating angioblasts, around day E8.5 (Dumont 1995; Rossant 2002; Thomas 2009). Thereafter, the *tie1* gene expression becomes restricted to endothelial precursor cells angioblasts, and later to the developing embryonic endothelium (Dumont 1995). The expression of Tie1 persists in quiescent adult endothelial cells at a reduced level. Higher Tie1 expression seen in the adult lung might reflect differences in the turnover rates of various endothelial cell populations. (Hobson 1984; Partanen 1999). The expression of Tie1 is further induced during physiological and pathological neovascularisation processes involved in ovarian-follicle maturation, wound healing, and tumor angiogenesis (Hobson 1984; Korhonen 1992).

The second Angiopoietin receptor, Tie2, is expressed by endothelial cells as well as hematopoietic cells (Thomas 2009). During embryonic development the *tie2* gene becomes transcriptionally activated already in hemangioblasts (Dumont 1995; Rossant 2002). Thereafter, the *tie2* gene expression becomes highly expressed in angioblasts, and later restricted to the developing embryonic endothelium. The expression of Tie2 persists in quiescent adult endothelium of all normal tissues, suggesting a role in maintenance of adult vasculature (Peters 2004). In the adult vasculature, Tie2 is upregulated and activated in areas of active angiogenesis like the ovaries and healing skin wounds (Partanen 1999).

Tie2 is also expressed by a subpopulation of hematopoietic stem cells and bone marrow osteoblasts (Dumont 1995). Tie2 is expressed by tumor cells e.g. Kaposi sarcoma cells and melanoma cells (Thomas 2009). Moreover Tie2 expression is upregulated during tumor angiogenesis including breast tumor, and non-small cell lung carcinomas (Peters 1998; Takahama 1999; Martin 2008). Of particular interest is the expression of the Tie2 receptor on a subpopulation of monocytes that seems to primarily account for the angiogenic activity of recruited tumor-associated macrophages (De Palma 2003; De Palma 2005; Augustin 2009).

Physiological roles of the Ang/Tie family members

Mice lacking the *Tie1* gene die between E13.5 and P1 due to loss of structural integrity of vascular endothelial cells, resulting in edema and hemorrhages (Sato 1995; Puri 1999). This mouse model suggests that Tie1 plays an important role in the regulation of the vessel integrity.

Deletion of the *Tie2* gene results in embryonic death between E9.5 and E10.5 due to vessel remodeling defects (Sato 1995) (Dumont 1994). Vessels are only poorly organized, have fewer branches and have reduced pericyte coverage. In another model, loss of Tie2 function leads to endothelial cell apoptosis which results in hemorrhages. These results suggest that the Ang/Tie system plays a key role during vessel remodeling, maturation and stabilization of cardiovascular system.

The double-knockout mice for Tie1 and Tie2 die, like Tie2-deficient mice, around E10.5 due to severe defects in the vascular system (Puri 1999). These results suggest that vasculogenesis proceeds normally without the Ang/Tie system, but Tie1 and Tie2 are essential later for maintaining the integrity of mature vessels.

Ang-1 deficient mice phenocopy the early embryonic lethal phenotype of Tie2-deficient mice (Suri 1996). Ang-1 deficient mice die between E11-E12.5 due to severe heart and vascular defects. The mice have growth-retarded hearts and an immature primary capillary plexus. Periendothelial cells appear not associated with endothelial cells (Augustin 2009).

Ang-2-deficient pups in the 129/J background are born alive; however, they die within 14 days after birth as a consequence of chylous ascites (Gale 2002; Augustin 2009). This rather mild phenotype provides the evidence that Ang-2 is dispensable for normal embryonic development and that Ang-2 may have an essential function in lymphangiogenesis. Ang-2-deficient mice show only minor vascular defects (Hackett 2002). Eye lens vessels regress shortly after birth in wild type, but not in Ang-2-deficient mice. This reflects a role of Ang-2 in vascular regression and vessel remodeling. The systemic Ang-2 overexpression phenotype is highly reminiscent of the phenotype of Ang-1- and Tie2-deficient mice (Maisonpierre 1997). This remark supports the hypothesis that Ang-1 acts in a stimulating, agonistic manner on Tie2, whereas, Ang-2 exerts antagonistic functions on Ang-1/Tie2 signaling (Thomas 2009).

Ang-2 has also been reported to have different effects dependent on the cytokine milieu (Kim 2000; Teichert-Kuliszewska 2001; Daly 2006). Ang-1 acts to stabilize vessels. Ang-2 can locally antagonize Ang-1-mediated Tie2 activation and therefore cause destabilization of vessels (Maisonpierre 1997). In the presence of both, Ang-2 and VEGF-A act together to induce angiogenesis (Rossant 2002). In the absence of VEGF-A, Ang-2 destabilizes vessels, leading to endothelial cell apoptosis, and thereby to vessel regression (Rossant 2002).

PDGF\PDGFR system

Another important controller of angiogenesis is the PDGF\PDGFR system, which drives the vessel remodeling and maturation.

Platelet derived growth factors (PDGFs) -A, -B, -C, and -D signal through their receptors PDGFRα and -β. PDGF-B/PDGFRβ signaling has been found to play a remarkable role during embryonic vascular development by the recruitment of pericytes to the newly formed blood vessels (Rossant 2002).

Endothelial cells secrete PDGF-B, which induces the proliferation and migration of vascular mural cells that express the PDGFRβ receptor (Armulik 2005). Deletion of either *Pdgfb* or *Pdgfrb* in mice leads to perinatal death caused by a reduced PC recruitment and coverage of microvessels (Lindahl 1997; Crosby 1998; Hellstrom 1999). Loss of pericytes leads to secondary effects, like endothelial hyperplasia, abnormal junctions and compensatory upregulation of VEGF-A, causing vascular leakage and hemorrhage (Hellstrom 2001).

1.4. The C-Myc proto-oncogene

The *myc* gene was originally identified as the transforming gene (*v-myc*) of the avian **my**elocytomatosis virus (MC29) (Kato 1992). The cellular homologue of the *v-myc* gene is the *c-myc* gene, discovered in chicken and found to be highly conserved in many species like human, mouse, rat, Drosophila (Dalla-Favera 1982; Vennstrom 1982; Stanton 1984; Hayashi 1987; Gallant 1996). Genes of the human Myc family have been found to be involved in numerous human neoplasias. The family of *myc* genes includes five members c-*myc*, B-*myc*, L-*myc*, N-*myc*, and s-*myc*; however, only c-*myc*, L-*myc*, and N-*myc* have neoplastic potential in humans (Schwab 1983; Nau 1985; Ingvarsson 1988; Sugiyama 1989).

1.4.1 Structure, function and transcriptional regulation of c-Myc

The c-Myc proto-oncogene encodes a nuclear phosphoprotein, which functions as a transcription factor (Pelengaris 2002a). The *c-myc* gene consists of three exons where exon 2 and 3 encode the c-Myc protein. The N terminus of the c-Myc protein contains three highly conserved elements among MYC family proteins, known as Myc Boxes I-III (Figure 6) (Pelengaris 2002a). They are required for the transactivation or transrepression of many target genes. The C terminus contains a dimerization motif, the basic-region/helix-loop-helix/leucine-zipper (BR/HLH/LZ) domain, which mediates heterotypic dimerization with other BR/HLH/LZ proteins (Patel 2004). Max was identified as an obligate partner protein of c-Myc. Although homodimers of c-Myc form at high concentration *in vitro*, all c-Myc proteins are found as heterodimers with Max *in vivo*. During dimer formation c-Myc appears to be the limiting, regulated component of the heterodimer, with its short mRNA and protein half lives (30min and 20min) (Gardner 2002). Max mRNA half life is 3hrs and its protein half life is >24hrs respectively (Gardner 2002). Max can form homodimers and it can also bind other BR/HLH/LZ Mad family proteins like Mad1, Mxi-1 (also known as Mad2), Mad3, Mad4 and Mnt (also known as Rox) (Nilsson 2003; Patel 2004; Adhikary 2005).

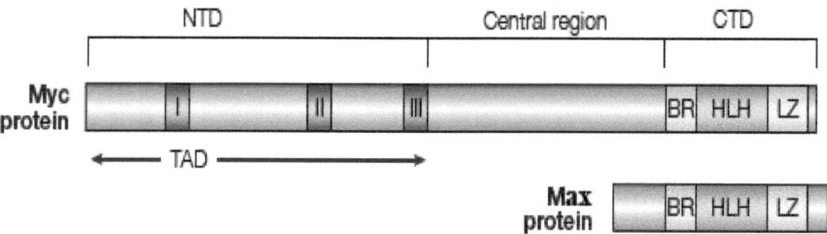

Figure 6. Domains of the human C-Myc protein
The carboxy-terminal domain (CTD) of human C-Myc protein harbours the basic-loop-helix (HLH) leucine zipper (LZ) motif for dimerization with its partner, Max, and subsequent DNA binding of Myc-Max heterodimers. The amino-terminal domain (NTD) harbors conserved Myc Boxes (MBI, MBII and III), which are essential for the transactivation of c-Myc target genes. TAD: transactivation domain. Modified from (Pelengaris 2002a).

Depending on the interactions with other proteins c-Myc can both activate and repress transcription of its target genes (Figure 7). The Myc-Max dimer activates transcription when bound directly to its consensus DNA recognition sequence (CACGTG, called E-box) and induces interactions of the Myc-boxes with transcriptional coactivators (such as TRRAP) (Pelengaris 2002a; Nilsson 2003). In most but not all the cases c-Myc activates transcription and repression of target genes through the Myc Box II.

The Mnt-Max or Mad-Max dimers actively mediate transcriptional silencing through direct protein-protein interactions (Nilsson 2003). Mad-Max heterodimers often accumulate in differentiating cells and are thought to silence proliferative genes that are activated by c-Myc during differentiation (Adhikary 2005).

Furthermore, the Myc-Max heterodimer can repress transcription by binding through the N-terminal region to other transcriptional activators such as Miz1 and thereby disrupting their function (Adhikary 2005).

In vivo there might be a regulated transition between an 'activating' and 'repressive' state of c-Myc depending on the physiological status of the cell (Adhikary 2005).

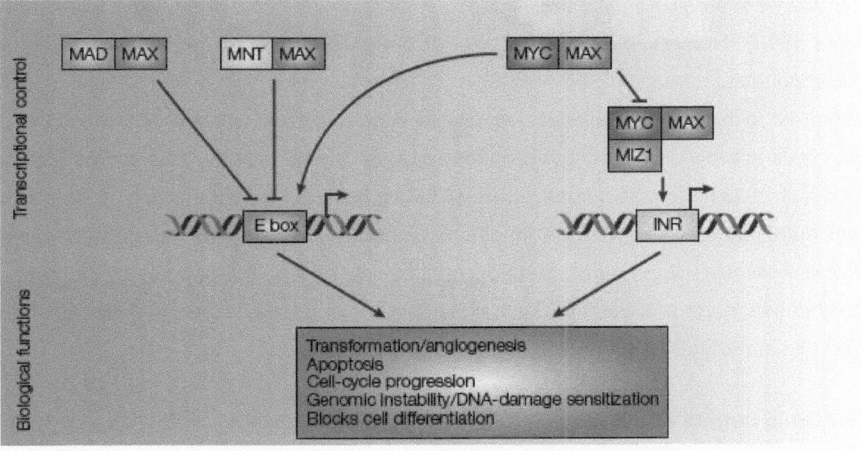

Figure 7. Transcriptional regulation by c-Myc
C-Myc binds DNA with its partner Max at the E-box sequences and either activates or represses transcription. Max also dimerizes with Mad or Mnt family proteins and mediates transcriptional repression of genes associated with E-box sequences. At certain promoters, c-Myc and Max interact with Miz1 to repress transcription. This generally occurs through a DNA element that overlaps the site where transcription begins, termed the initiator element (INR). Together, c-Myc and its partners regulate many important biological processes, including transformation, apoptosis, cell-cycle progression, genomic stability, the DNA-damage response, differentiation, and angiogenesis. Modified from (Patel 2004).

1.4.2 C-Myc and cell proliferation

C-Myc is an essential part of the normal cell proliferative machinery (Eisenman 2001). The resting cell (G0) normally expresses little *c-myc*, whereas cells stimulated by growth factors dramatically increase *c-myc* expression as an immediate early response gene (Gardner 2002). Its expression is activated by a variety of mitogens (growth factors, extracellular contacts or internal clocks, such as the cell cycle) during the G0-G1 phase of the cell cycle. C-Myc expression persists into cell cycle, this basal level of expression is dependent upon continuous growth factor stimulation. Then c-Myc expression returns to its basal quiescent state in resultant resting daughter cells (Gardner 2002). In adult tissues, *c-myc* expression is generally correlated with the proliferative rate of the tissue (Lemaitre 1996).

Normal embryonic development requires regulated expression of *c-myc*. Mice with homologous deletion of c-Myc was lethal between E9.5 and E10.5 of gestation. The embryos were smaller in size and retarded in development compared with their littermates (Davis 1993). This result indicates a role for c-Myc in regulating cell division during early embryogenesis (Lemaitre 1996).

In contrast to the tightly regulated *c-myc* gene in normal cells, which only express the gene when cells actively divide, cancer cells may express the gene in an uncontrolled fashion as the result of genetic aberrations. In most tumor cells, c-Myc expression is deregulated, often markedly elevated, and no longer dependent upon mitogenic signaling (Pelengaris 2002a). Activation of the *c-myc* gene occurs in several ways, for example by chromosomal translocation, gene amplification, retroviral insertion, or increased *c-myc* gene transcription or protein stability (Pelengaris 2002a).

1.4.3 Gene targets of c-Myc

After its activation, c-Myc functions as a master transcriptional regulator of a wide number of target genes that execute the cellular response. A surprising new finding is that *in vivo* c-Myc is bound to approximately 25.000 binding sites in the human genome (Adhikary 2005). Therefore, c-Myc regulates a significant proportion of all genes in an organism. The many target genes raise the question: which of these are the crucial targets of c-Myc? Several genes have been identified that function downstream of c-Myc in specific biochemical pathways.

c-Myc influences cell-cycle progression – C-Myc promotes cell cycle progression after mitogen or serum stimulation (Pelengaris 2002a). Several important studies could show how c-Myc activates or repress target genes that are involved in cell–cycle progression. C-

Myc induces cyclin-D-CDK4 and cyclin-E-CDK2 activity, which is an essential event in Myc-induced G1-S progression (Bouchard 1999; Coller 2000; Hermeking 2000). On the other hand Myc-Max heterodimer interacts with Miz1 transcription factor and inactivates INK4B and WAF1, CDK inhibitors (Herold 2002).

c-Myc inhibits cell differentiation - Numerous studies have highlighted the importance of MYC\MAX\MAD network in regulating cell proliferation and differentiation (Foley 1999; Grandori 2000; Pelengaris 2002a). MAD\MXI1 proteins heterodimerize with MAX and subsequently repress transcription by recruiting a chromatin-modifying corepressor complex to E-box sites on the same target genes as MYC-MAX, such that the MYC-MAX complex can no longer activate its target genes (Pelengaris 2002a).

c-Myc induces angiogenesis - the induction of VEGF-A expression contributes to the role of c-Myc in angiogenesis (Brandvold 2000; Okajima 2000; Baudino 2002). C-Myc has also been shown to be angiogenic through its repression of the angiogenesis inhibitor, thrombospondin-1 (Janz 2000; Ngo 2000; Pelengaris 2002a).

c-Myc induces tumorigenesis - the crucial pathways through which c-Myc exerts tumorigenic effects are not well defined. An important suggestion has been that cellular transformation induced by c-Myc depends on the deregulation of specific metabolic pathways (Adhikary 2005). For example, c-Myc transcriptional activates lactate dehydrogenase A expression (Shim 1997). C-Myc upregulates metabolic enzymes such as serine-hydroxymethyl-transferase (SHMT) and ornithine decarboxylase (ODC) (Nikiforov 2002; Bello-Fernandez 1993).

c-Myc induces apoptosis – c-Myc triggers rapid apoptosis through at least two distinct pathways (Figure 8). First, c-Myc induces the expression of $p19^{Arf}$, an inhibitor of MDM2, which leads to the stabilization of p53 (Zindy 1998). However, the $p19^{Arf}$ gene does not seem to be a direct c-Myc target, and the mechanism through which Myc activates $p19^{Arf}$ expression is unclear. Second, c-Myc promotes the release of cytochrome c from mitochondria and this is required for Bax oligomerization and function (Juin 1999; Mitchell 2000; Soucie 2001). This pathway is mediated partly by BIM, which is probably induced by c-Myc in an indirect fashion (Egle 2004). Furthermore, repression of the anti-apoptotic Bcl-X_L and Bcl2 proteins by c-Myc contributes to the release of cytochrome c (Eischen 2001). Deregulated expression of c-Myc might also induces apoptosis through the induction of DNA damage (Seoane 2002; Karlsson 2003a; Karlsson 2003b).

Myc-induced genomic instability - might also contribute to tumorigenesis (Vafa 2002). Several conditional knockout systems support the view that c-Myc facilitates the emergence and survival of cells that have sustained additional oncogenic mutations (Felsher 1999b; D'Cruz 2001; Shachaf 2004).

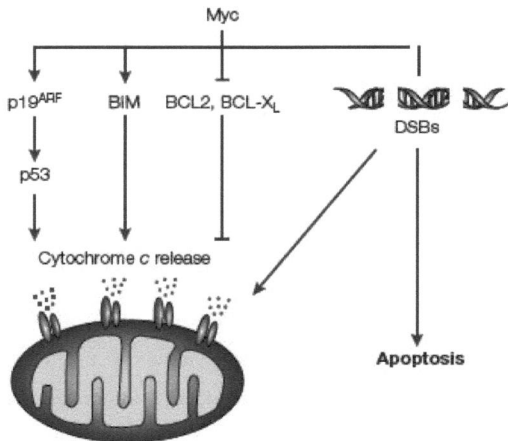

Figure 8. Pathways that are implicated in c-Myc-induced apoptosis
Several pathways have been implicated in Myc-induced apoptosis. First, c-Myc induces the expression of p19^{Arf}, which stabilizes p53; the induction of p19^{Arf} is probably indirect. Second, c-Myc induces the pro-apoptotic BIM protein, and third Myc blocks the expression of the anti-apoptotic factors Bcl2 and Bcl-X$_L$, which leads to the release of cytochrome c in a p53-independent manner. Another potential mechanism of c-Myc-induced apoptosis is the ability to cause DNA damage, DNA double-strand breaks (DSBs) and to block DNA-damage-dependent cell-cycle arrest. Figure adopted from (Adhikary 2005).

1.5. C-Myc in vascular development

Davis et al. described first the systemic c-Myc knock out mice (Davis 1993). They observed that homologous recombination of c-Myc was lethal between E9.5 and E10.5. Pathologic abnormalities include malformations of the heart and neural tube. The heart abnormality consisted of enlargement and a dilated fluid-filled pericardium. Yolk sac circulation was impaired, too, in the homozygotes embryos. The embryos were smaller in size and retarded in development as compared to their littermates. These results cannot exclude a role for c-Myc in regulating cell division during early embryogenesis. However some compensatory mechanism might occur in these embryos to replace c-Myc function, possibly carried out by another member of the Myc family (Lemaitre 1996).

Further studies of the c-Myc -/- embryos revealed a crucial role for c-Myc in vascular and hematopoietic development (Baudino 2002). C-Myc deficiency results in a compromised endothelial and erythroid cell number, abnormal yolk sac development with less hierarchical organized branches and a lack of clearly defined vascular structures. Baudino and colleagues described that c-Myc could function as a master regulator of angiogenic factors during vascular development (Baudino 2002). The developmental defects are associated with a requirement for c-Myc for the expression of VEGF-A, and other angiogenic factors like Ang-2, Ang-1 and thrombospondin-1. However, in these mice c-Myc was missing in all cell types so the specific contribution of c-Myc to endothelial cell physiology could not be evaluated.

THE AIMS OF THE STUDY

Previous work had shown that c-Myc is required for adequate vasculogenesis and angiogenesis during tumor formation as well as for embryonic development.

To further investigate the contribution of c-Myc to these processes, we conditionally expressed c-Myc in adult and embryonic endothelial cells using a tetracycline-regulated tet-off system.

The first part of this study elucidates the role of c-Myc in the adult vascular system. The vascular endothelial cell-specific c-Myc overexpression induced pathological malformations were analyzed. *In vivo* tumor growth upon oncogene deactivation was investigated. Tumor cell lines were established and characterized.

The second part of the work was dedicated to the role of c-Myc in endothelial cells during embryonic development. The effects of c-Myc overexpression in developing embryonic blood and lymphatic vascular system were characterized. C-Myc target genes were analyzed at a molecular level.

2. MATERIALS AND METHODS

2.1. Transgenic mice

The Tie2-tTA mouse line on a Bl6 background expresses the tetracycline-transactivator under the control of the Tie2 endothelial cell specific promoter (Deutsch 2008). The tetO-Myc NMRI outbread mouse line contains the human c-Myc cDNA and a luciferase reporter gene under the control of the tetracycline responsive bidirectional promoter (tetO) (Marinkovic 2004). In double transgenic animals Tie2-tTA / tetO-Myc (Tie2-tTA/tetO-Myc), tTA mediates the transcription of the luciferase reporter gene and simultaneously the c-Myc protooncogene expressed from the bi-directional promoter. The transcription of both genes is repressed in the presence of doxycycline (Deutsch 2008).

In general the tet-off system consists of two transgenic constructs (Gossen 1995). In our case, one of the mouse lines expresses the tetracycline-transactivator (tTA), which is a fusion protein between the tet repressor of the Tn10 tetracycline resistance operon from Escherichia coli and the C-terminal part (transactivating domain) of the VP16 from the Herpes simplex virus. In the other transgenic mouse line, the transgene is expressed from a tetracycline responsive minimal promoter (tetO). In double transgenic animals, the tTA mediates the transcription of the transgene. Presence of tetracycline prevents binding of tTA to the tetO promoter and results in an abrogation of the transgene transcription.

F1 embryos were used for experimental investigations deriving from intercrosses between tetO-Myc (NMRI outbred) and Tie2-tTA (C57BL/6 background) mouse lines. The age of embryos used for analysis was determined via vaginal plug control on day E0.5. To suppress expression of c-Myc transgene during embryonic development, doxycycline was given in the drinking water of pregnant mothers at a concentration of 2 mg/ml.

Adult mice were housed in the animal facility associated with the Institute for Laser Technology in Medicine and Measurement Technique (ILM) in open cages. In this type of cages the air exchanges freely across the cage top thus the cage-to-cage transmission of infectious agents is important (Fox 2007). These animals were transferred due to *in vitro* fertilisation in N26O mouse facility and were kept in individually ventilated cages (IVC). In IVCs filtered air is delivered under pressure to the individual cages. The cage-to-cage transmission of infectious agents is slowed or eliminated. Thus the microbiological status of individual cages is maintained (Fox 2007).

2.2. Animal experiments

2.2.1 Gross embryonic pathology

Pregnant mothers were sacrificed at different time points depending on the age of their embryos. Embryos were collected in ice cold PBS and dissect away extraembryonic membranes. Embryos were checked for morphological changes on the whole body. Double transgenic embryos showed clear edema and/or hemorrhages on their body. They were compared to the corresponding controls. Photographs were taken using a Leica microscope (DMIRB/E) and OpenLab software. Further analyses were made using histological and flow cytometry techniques.

2.2.2 Gross pathology of adult animals

Tumor bearing mice were sacrificed and examined for morphological changes in their body. The morphology of analyzed organs of tumor-affected mice was compared to the corresponding controls. Further analyses were made using histological techniques. Photographs were taken using a Canon camera.

2.2.3 Doxycycline treatment of tumor bearing adult mice

To abrogate c-Myc expression in tumor bearing mice, mice were treated with 2 mg/ml doxycycline dissolved in the drinking water for 4 weeks long. Tumor regression was followed over the time.

2.2.4 Tumor cell line transplantation

Rag2 -/- mice were inoculated subcutaneously in a maximal volume of 150 µL 0,9% NaCl solution with 2×10^7 cells. After the tumor volume reached the 0,5 cm in diameter, a control group of animals were treated with 2 mg/ml doxycyline in the drinking water. The experimental group of animals was not treated with doxycycline. The growing of the tumor mass was followed up by measuring the tumor volume with a caliper rule.

2.3. Genotyping of transgenic mice

2.3.1 Genomic DNA isolation

As material for DNA isolation from adult animals a small piece of mouse tail was used. Embryo genotyping was performed with a small piece of embryonic tissue. Tissue was incubated in 750 µl of 1xTNES buffer with 0.45 mg/ml of Proteinase K at 56 °C for 12 hours. 250 µl 6M NaCl were mixed to the solution and centrifuged at maximum speed for 10 minutes. Supernatant was transferred into a clean tube and filled with equal volume of Isopropanol. DNA was precipitated by shaking the tube and centrifuged at maximum speed for 10 min. Pellet was washed with 70 % EtOH and dried at room temperature for 10-20 minutes. DNA was dissolved in 250 µl of distilled water.

2.3.2 Polymerase Chain Reaction (PCR)

DNA was isolated as previously described and PCR mix was prepared by following scheme:

6 µl	5x PCR Buffer
3 µl	2 mM dNTPs
1 µl	10 pmol/µl primer 1
1 µl	10 pmol/µl primer 2
0,2 µl	5 U/µl Taq Polymerase
2 µl	genomic DNA
add	ddH$_2$O to 30µl

primer for TetO-Myc mice:
myc: 5'- GGG GAG GAC TCC GTC GAG G -3'
pBi5: 5'- GGT ACC CGG GGA TCC TCT AGT CAG -3'
annealing temperature: 63°C

primers for Tie2-tTA mice:
tTA1: 5'- GAT AGG CAC CAT ACT CAC TTT TGC CC -3'
tTA2: 5'- CGA TAG CTT GTC GTA ATA ATG GCG GC -3'
annealing temperature: 63°C

PCR conditions:

1x	3 min	94°C
32x	40 sec	annealing temperature
	60 sec	72°C
	40 sec	94°C
1x	1 min	annealing temperature
	10 min	72°C

Obtained PCR products were separated by electrophoresis using 1% agarose gel and TAE buffer and visualized with 4 µg/ml of ethidium bromide under UV-light.

2.4. Detection of gene expression

2.4.1 Reporter gene detection

Luciferase activity in protein extracts

Tissue was homogenized in luciferase extraction buffer (described in Materials section). Protein extract was diluted in luciferase buffer, which contains Luciferin the substrate for the enzyme. Firefly luciferase enzyme activity was measured as emitted bioluminescence using a luminometer (Lumat LB 9570, Berthold Technologies). Protein concentration was measured using Bradford solution. Afterwards lightunits of luciferase activity were normalized to the total protein concentration.

Visualization of luciferase expression on tissue samples

20 µm thin whole embryo cryosections were placed on normal glass slides and after drying the border was signed using pap-pen. Tissue was covered with 100µl of luciferase buffer. Light emission was measured in a Dark Box system (Luminescent Image Analyzer LAS-1000) data was analysed with Openlab software.

2.4.2 Analysis of mRNA expression levels

RNA isolation

RNA was isolated using High pure RNA isolation kit (Roche), according manufacturer's instruction. Briefly, tissue powders were lysed with a strong denaturating buffer containing guanidine hydrochloride. In the presence of chaotropic salt (guanidium HCl) RNA binds selectively to a glass fiber fleece. Residual DNA was digested by DNase I. After several "wash-and-spin" steps, pure RNA was eluated by H_2O and directly used for cDNA synthesis or stored at -20°C.

cDNA synthesis

RNA was reverse transcribed using MuMLV reverse transcriptase kit (Promega) according to manufacturer's instruction. Total RNA was denaturated for 10 minutes at 70°C in the presence of random hexamer primers and RNase inhibitor and than rapidly cooled down on ice to enable binding of primers. After this steps 5x buffer, DTT, dNTPs and MuMLV reverse transcriptase were added. Reaction was performed at 37°C for 2 hours and then stopped by inactivation of enzyme for 5 minutes at 70°C. Obtained cDNA was used for PCR reaction or stored at -20°C.

RT-PCR

cDNA obtained by described protocol was diluted 1:10 and subjected to the PCR reaction using primers listed in Table 2. As housekeeping gene β-actin was used.

For all PCRs following mix was prepared

1 µl	cDNA
3 µl	10x PCR Buffer
3 µl	2 mM dNTPs
1 µl	10 pmol/µl forward primer
1 µl	10 pmol/µl reverse primer
0,2 µl	5 U/µl Taq Polymerase

add ddH2O to 30µl

PCR conditions:

1x	3 min	94°C
32x	40 sec	annealing temperature
	60 sec	72°C
	40 sec	94°C
1x	1 min	annealing temperature
	10 min	72°C

Obtained PCR products were separated by electrophoresis using 1% agarose gel and TAE buffer and visualized with 4 µg/ml of ethidium bromide under UV-light.

2.4.3 Quantitative real-time PCR analysis

Total RNA was isolated from sorted embryonic endothelial cells using PicoPure RNA Isolation Kit (Arcturus Bioscience). cDNA preparation protocol was described before. Real-time quantitative PCR was performed using FastStart SYBR Green PCR Kit (Roche) following manufacturer protocol. A P value <0.05 was considered statistically significant. Primers sequences used for quantitative RT-PCR are listed in Table 3.

2.5. Protein analysis

2.5.1 Protein concentration measurement by Bradford method

1 µl of cell extracts (made with RIPA, DIGNAM C or ELISA sample buffer) was mixed with 100 µl of 150 mM NaCl and 1 ml of Bradford reagent. Optical density of formed protein/coomassie-briliantblue complex was measured in photometer at wavelengths of 595 nm. Protein concentration in extracts was estimated according standard curve with known concentrations of bovine serum albumin. Protein extracts were shock-freezed in liquid nitrogen and stored at -80°C.

Protein isolation and separation by SDS-PAGE

Embryonic lungs were washed with phosphate-buffered saline (PBS), lysed on ice in 250 µL of 2 x concentrated Laemmli-Urea sample buffer. Proteins were denatured by incubation at 95°C for 5 minutes following by a brief spin down. 25 µl of lysates were separated on a 10% SDS-polyacrylamide gel by electrophoresis. SDS-polyacrylamide gels used in this study were composed of 5% stacking ("upper") and 10% resolving ("lower") gels made with appropriate buffers described in Materials. Electrophoresis was performed in SDS-running buffer for 45 min by 150V.

Adult mice tissues were washed with phosphate-buffered saline (PBS) following protein extracts preparation in RIPA or DIGNAM C protein extraktion buffer using Turax. Protein concentration was measured by Bradford method. By this step extracts can be stored at -80 °C or used for Western blotting immediatelly. 4x Laemmli buffer was given to 50 µg of protein extracts following denaturation by incubation at 95°C for 10 minutes. Samples were spun down briefly. Denaturated extracts were loaded and separated on 10% SDS-polyacrylamide gel by electrophoresis for 45 min by 150V in SDS-running buffer.

2.5.2 Western immunoblot

Proteins separated on SDS-polyacrylamide gel were blotted onto PVDF membrane for further analysis. Membrane was activated by methanol and proteins were transferred from the gel to the membrane by "semi-dry blot" principle in 0,5 A electric field for 2 hours at 4°C in semi-dry transfer buffer. Membrane blocking and incubation with antibodies involved standard procedures of 1 hour incubation followed by three times washing with PBS containing 0,1% Tween20. To prevent non-specific binding of antibody, membrane was incubated in the solution of 5% non-fat milk in PBS and then probed with an antibody diluted in PBS containing 0,1% Tween20 and 5% skim milk. Binding of primary antibody was revealed using HRP-conjugated appropriate secondary antibody and ECL Western Blotting Detection Reagents. Antibodies used in this study are listed in Table 1.

2.5.3 VEGF-A ELISA

Embryonic lungs were snap frozen and kept at -80°C until analysis. Lungs were subsequently lysed and homogenized in ELISA Buffer. Total protein concentration was measured by Bradford method. VEGF-A ELISA detection kit (R&D Systems) was used to quantify VEGF-A protein concentration in lung tissue lysates. VEGF-A sandwich ELISA was performed following manufacturer protocol. The 96-well plates were read in a SpectraMax 190 microplate reader. VEGF-A concentration was normalized against the total amount of protein in the sample.

2.5.4 Total MMP-9 ELISA

Protein homogenates from embryonic lungs was prepared as before by VEGF-A ELISA measurement. Total protein concentration was measured by Bradford method. MMP-9 ELISA detection kit (R&D Systems) was used to quantify total MMP-9 protein from tissue lysates. MMP-9 sandwich ELISA was performed following manufacturer protocol. ELISA plates were read in a SpectraMax 190 microplate reader. Total MMP-9 concentration was normalized against the total amount of protein in the sample.

2.6. Flow cytometry analysis

Flow cytometry is the technological process that allows for the individual measurements of cell fluorescence and light scattering. This information can be used to individually sort or separate subpopulations of cells. The process of cell sorting is performed at rates of thousands of cells per second. Flow cytometry is performed on a single cell suspension.

2.6.1 Isolation of primary embryonic endothelial cells

Isolation of endothelial cells from embryonic murine tissue was performed according to the protocol described elsewhere (Marelli-Berg 2000), with some modifications. In brief, embryonic tissue was dissected into 2-mm3 blocks, washed in PBS supplemented with 2% FCS. Single cell suspension was achieved due to enzymatic digestion of the tissue peaces. Diced tissue was incubated with 1 mg/ml Collagenase/Dispase (Roche) and 0,25 µg/ml DNase I (Qiagen) for maximal 20 min at 37 °C in a humid incubator and resuspended every 5 min. Digested tissue were passed through a cell strainer (74 µm mesh, Costar) to remove undigested blocks. Cells were washed in PBS supplemented with 2% FCS and centrifuged at 1700 rpm, 8 min. Estimation of cell numbers was performed by Trypan-blue exclusion. Cell pellet was used for further cytometry analyses.

2.6.2 Detection of apoptosis and proliferation by flow cytometry

For detection of apoptosis and proliferation of embryonic endothelial cells following kits were used: PE-conjugated monoclonal active caspase-3 antibody apoptosis kit and PE-conjugated Ki67 mouse a-human monoclonal antibody set from BD (Table 1). Staining procedure was performed as described in manufacturer's manual. Single cell suspensions isolated from embryonic tissues were incubated with the combination of anti-CD31/Pecam-1-FITC and appropriate antibody for apoptosis or proliferation. Cells were resuspended in 500 µl of FACS buffer (PBS supplemented with 2% FCS, 2 mM EDTA, 0,25 µg/ml DNase I), filtered through a cell strainer and analyzed. Analyses were performed with a four-colour flow cytometer (FACSCalibur) and CellQuest software.

2.6.3 Fluorescence Activated Cell Sorting (FACS)

Embryonic endothelial cells were sorted with the aim to get a clean endothelial cell population. Single cell suspensions isolated from embryonic tissues were incubated to block FcγII/III receptors with 1:100 dilution of anti-CD16/CD32 antibody in PBS containing 2% FCS for 30 minutes at +4°C. The cells were incubated for 30 min, on ice in the dark with the combination of anti-CD31/Pecam-1-FITC (1:100) and anti-CD105/Endoglin-biotin (1:100) antibodies diluted in PBS/2%FCS/0,25 µg/ml DNase I. After washing with PBS/2%FCS secondary antibody staining was performed with streptavidin-PE (0.65:100) diluted in PBS/2%FCS/0,25 µg/ml DNase I for 30 min, on ice in the dark. After following washing, cells were resuspended in 500 µl of FACS buffer (PBS supplemented with 2% FCS, 2mM EDTA, 0,25 µg/ml DNase I) and filtered through a cell strainer. Sorting was performed on FACS Aria cell sorter instrument. Sorted embryonic endothelial cells were used for further analysis.

2.7. Histological analyses

2.7.1 Immunohistochemistry on tissue sections

Antibodies used in immunohistochemistry studies are listed in Table 1.

Tissue embedding and sectioning

For histological analyses tumor tissue or whole embryos were embedded in Tissue-Tec O.C.T. compound in cryomolds, cooled down slowly in dry-ice-Methylbutan mixture. Frozen samples were stored at -80 °C. 5 µm thin sections were prepared in cryostat at -20°C, air dried overnight at room temperature and stored at -80°C or stained immediately.

Hematoxylin-Eosin staining

Sections were fixed in -20°C cold aceton for 10 minutes and air-dried for 15 minutes. Sections were kept 3 minutes in Hemalaun solution, which is a basic dye containing hematoxylin, following 3 min washing under running tap water. Staining with 0.1% Eosin solution was performed 20 sec long. Tissue samples were dehydrated in ethanol, cleared in xylol and mounted with Eukitt (Merck). Slides were analyzed using a Leica DMIRB/E microscope and OpenLab software.

Hematoxylin stains blue the cell nuclei and basophilic structures (structures taking up the basic dyes, these are nucleic acid containing structures like nucleus and ribosomes). Cytoplasm and eosinophilic / acidophilic structures, like extracellular proteins are red colored with 0.1% Eosin solution (acidic dye).

Immunohistochemistry

For the analysis of LYVE-1 and CD31/Pecam-1 protein expression in the embryos, serial transverse cryosections were prepared, as described before. Dry sections were fixed in -20°C aceton for 10 minutes. Samples were stained with 1:100 dilution of goat anti-mouse LYVE-1 antibody, raised against the extracellular domain, followed by staining with a secondary rat anti-goat HRP-conjugated antibody. Samples incubated with rat anti-mouse CD31/Pecam-1 antibody, were stained with goat anti-rat HRP-conjugated secondary antibody. Binding of HRP-conjugated antibody was visualized using AEC compound. 4 mg of AEC was dissolved in 0,5 ml DMF (dimethylformamid) and than mixed with 2,1 ml of 0,1 M acetic acid and 7,4 ml of 0,1 M Na-acetate. Solution was filtered through 0,45 µm filter and 5 µl of H_2O_2 was added just before use. Color development was allowed maximal 30 min long. Color reaction was stop rinsing the slides in tap water. Slides were counterstained with hematoxylin, washed with tap water and mounted with Aquatex (Merck) mounting media. Analyses were performed using a Leica microscope (DMIRB/E) and OpenLab software.

Immunofluorescence

Immunofluorescence double staining for LYVE-1 and CD31/Pecam-1 was performed on transverse embryonic cryosections. Tissue sections were fixed in -20°C aceton for 10 minutes. Slides were incubated 45 min with blocking buffer for blocking unspecific bindings to the Fc-receptors: staining buffer (PBS containing 0,1 % BSA) plus 1:100 diluted anti-CD16/CD32 antibody. Blocking buffer was removed following antibody staining. Antibody staining was performed for 45 min in staining buffer with 1:100 dilution of both LYVE-1 and CD31/Pecam-1 antibodies. Sections were washed 3 times 3 minutes with PBS. Fluorochrome conjugated secondary antibody incubation take 45 min with 1:500 dilution of goat anti-rat Alexa 488 and goat a-rabbit Texasred antibodies. After washing sections were mounted with Mowiol + 0,1 μg/mL DAPI. Sections were analyzed on ZEISS ApoTome microscope and AxioVision software. Antibodies used in this study are listed in Table1.

2.7.2 Immunohistochemistry on tissue peaces: Whole-mount analyses

Whole-mount immunohistochemistry allows visualization of antigens (usually proteins) within an embryo or within a peace of tissue like an inner organ or skin. Briefly, after fixation tissue samples were permeabilized and blocked in one step, following antibody staining. Typically, a primary antibody binds specifically to an antigen; then a secondary antibody conjugated to a flourochrome or enzyme (HRP) binds to the heavy chain constant region of the primary antibody with specificity to the species of origin of the primary antibody (e.g. anti-rabbit conjugated to Rhodamine). By HRP stained samples the signal was visualized by a developing reaction using DAB. Subsequently the whole specimen was placed on a slide just before microscopic examination.

Whole-mount immunohistochemistry

Freshly isolated embryos were fixed in 4% PFA overnight at +4°C. Tissues were washed twice with PBS for 10 min at +4°C. Dehydrate embryos in methanol series: 25% MeOH in PBS, 50% MeOH, 75% MeOH and 100% MeOH (twice) each for 15 minutes at room temperature. To bleach the tissue and block endogenous peroxidase, embryos were incubated with 5% H_2O_2 in MeOH for 4-5 hours. Bleaching was stopped by rinsing the embryos with 100% MeOH twice. At this point tissues can be stored in MeOH at -20°C for at least a few weeks. Subsequently rehydrate tissue through 75% MeOH, 50% MeOH, 25% MeOH and PBS (twice) each for 15 minutes at room temperature. Blocking was performed incubating the tissue in PBSMT for 2 hours at +4°C. Antibody staining was performed with overnight incubation at +4°C in PBSMT diluted primary antibody, rat anti-mouse CD31/Pecam-1 1:300 dilution, by gently rocking the samples. During the next day samples were washed five times with PBSMT for 1 hour at +4°C, following secondary antibody staining, goat anti-rat HRP 1:200 dilution, overnight. During the next day samples were washed 5 times with PBSMT every hour and once with PBT for 20 minutes at room temperature. Finally developing solution was prepared: 10 mg DAB was dissolved in 10 ml PBT, after DAB dissolved 0.17g NiCl2 was given and dissolved too. Solution was filtered and 3 parts of this concentrated developing solution was mixed with 7 parts of PBT (final concentration is 0.3 mg/ml DAB, 0.5% NiCl2 in PBT). Embryos were incubated in the diluted developing solution 20 minutes at room temperature in the dark. Then H2O2 was added to a 0.03% final concentration (1μl of 30% H2O2 for 1 ml developing solution) and immediately mixed gently. Color was developed within 5-10 minutes, to stop the reaction, embryos were rinsed in PBT twice for 5 minutes then in PBS twice for 5 minutes at room temperature. Subsequently tissues were postfixed, rocking them in 2% PFA / 0.1% glutaraldehyde / PBS at +4°C overnight. On the next day tissues were rinsed in PBS three times 5 minutes at room temperature. In order to take whole embryo photographs, tissue was equilibrated in 50% glycerol at room temperature for 1 hour then in 70% glycerol for 1 hour. Pictures were taken in 70% glycerol using a Leica microscope (DMIRB/E) and OpenLab software.

After analyses embryos can be rinsed in PBS several time to remove glycerol then dehydrate through MeOH series for paraffin embedding.

Antibodies used in this study are listed in Table 1.

Whole-mount immunofluorescence staining

Fresh embryos were fixed in 4% PFA at +4°C over night. In the next morning embryos were washed twice with PBS for 10 min at +4°C, subsequently unspecific bindings were blocked with incubation in PBSMT for 2 hours at +4°C. Primary antibody staining was performed with rat anti-mouse CD31/Pecam-1 antibody 1:300 diluted in PBSMT overnight at +4°C. During the next day samples were washed 4-5 times with PBT for 1 hour at +4°C. Afterwards fluorochrome conjugated secondary antibody staining was performed overnight at +4°C with goat anti-rat Alexa Fluor 488 antibody 1:500 diluted in PBT. During the next day samples were washed 4-5 times with PBT for 1 hour at +4°C to wash out the unbounded secondary antibody from the tissue. Stained blood vessels were analyzed on

ZEISS ApoTome microscope using AxioVision software. Antibodies used in this study are listed in Table 1.

Whole-mount double immunofluorescence staining

For study the embryonic blood vessel structure double immunofluorescence labeling was performed on embryonic back tissue pieces either with CD31/Pecam-1 antibody for endothelial cells and NG2 proteoglycan for pericytes or with CD31/Pecam-1 antibody for endothelial cells and Laminin antibody to stain basement membrane or with CD31/Pecam-1 antibody for endothelial cells and α-SMA antibody for smooth muscle cells.

Freshly isolated embryonic tissue pieces from the back were fixed in 4% PFA for 2 hours at +4°C. Tissues were washed twice with PBS for 10 min at +4°C, subsequently unspecific bindings were blocked with incubation in PBSMT for 2 hours at +4°C. Primary antibody staining was performed either with rat anti-mouse CD31/Pecam-1 antibody (1:300) and NG2 (1:200) or with rat anti-mouse CD31/Pecam-1 (1:300) and Laminin (1:200) or with CD31/Pecam-1 (1:300) and α-SMA (1:200) in PBSMT overnight at +4°C. During the next day samples were washed 4-5 times with PBT for 1 hour at +4°C. Afterwards fluorochrome conjugated secondary antibody staining was performed with goat anti-rat Alexa 488 and goat anti-rabbit Texasred antibodies 1:500 diluted in PBT overnight at +4°C. During the next day samples were washed 4-5 times with PBT for 1 hour at +4°C to wash out the unbounded secondary antibody from the tissue. Stained blood vessels were analyzed on Zeiss ApoTome microscope with AxioVision software. Antibodies used in this study are listed in Table 1.

2.7.3 Electron Microscopy

For electron microscopy, different organs (heart, intestine, liver, kidney, brain, skin of different locations like limbs, tail and back) of E14.5 old single or double-transgenic embryos were fixed for 4 hours in 2.5% glutaraldehyde (Paesel and Lorei) in 0.1 M cacodylate buffer (pH 7.4). Specimens were washed in pure cacodylate buffer, postfixed overnight in 1% OsO4 in cacodylate buffer for 1h, dehydrated in ascending series of ethanol and propyleneoxide, bloc-stained in uranyl-acetate for 4 h and flat-embedded in Araldite (Serva). Using an ultramicrotome (Ultracut, Leica), semi-(1µm) and ultrathin sections (50 nm) were cut. Ultrathin sections were stained with lead citrate, mounted on copper grids and finally analysed with a Zeiss EM 10 electron microscope.

2.8. Quantitative analysis of immunohistological stainings

2.8.1 Staining quantification

Lymphatic vasculature in the skin and jugular lymph sac area was analyzed from double fluorescence immunostained sagittal embryonic cryosections by E14.5 animals. Fluorescence images were taken from sections stained with LYVE-1 (green) and CD31/PECAM-1 (red) antibodies. The single color images were converted to 8-bit grayscale using Adobe PhotoShop software. Then images were transformed into binary images by automatic thresholding using ImageJ software in order to classify LYVE-1 and CD31/PECAM-1 positive area, respectively (Ridler 1978). As a quantification of LYVE-1 staining, the LYVE-1 positive area divided by the CD31/PECAM-1 area was estimated (Tammela 2005).

2.8.2 Quantitative analysis of dermal vascular architecture

For the quantitive analysis of vessel morphology the blood vessels were regarded as a random network of line segments neglecting their width and curvature.

In order to transform the grayscale images of the blood vessels into a network of line segments, the images were first skeletonized. Therefore the varying brightness of the images was corrected and the images were smoothed by anisotropic diffusion. Then the smoothed images were transformed into binary images by thresholding. In the next step a distance transformation was applied and finally the transformed images were skeletonized using the watershed transformation. For details about the used image processing techniques see (Soille 1999). In the last step the skeleton was transformed into a network consisting of line segments as described in (Beil 2005). An image together with the network of line segments is displayed in Figure 9. . Based on the network of line segments different statistical characteristics describing the spatial-geometric structure like the mean number of branching points and line segments both measured per unit area and in addition the mean length of the segments were estimated. For the estimation procedure a sampling window W with area |W| was chosen. Then the number of branching points and segments in W was counted and divided by |W|. A segment was counted if its lexicographically smallest point was inside W. Furthermore, the mean length of the segments in each images was estimated using again the line segments whose lexicographically smallest point was inside W. All these estimators are unbiased (Baddeley 2004).

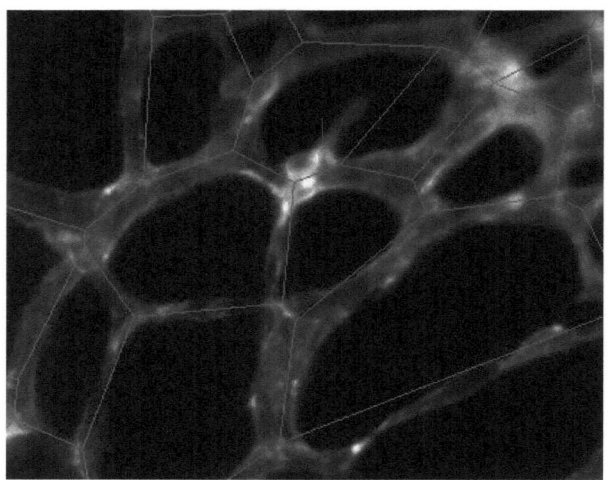

Figure 9. A grayscale image of blood vessels together with the constructed network of line segments.

2.9. Materials

2.9.1 General chemicals

All general chemicals i.e. salts, acids etc. were from the following companies: Sigma-Aldrich, Gibco, Merck, Fluka, AppliChem GmbH, Roth.

2.9.2 General buffers and solutions

10x PBS   1 l (10x)

NaCl	137 mM	87,65 g
KCl	2,6 mM	2 g
NaH_2PO_4	1,8 mM	2,4 g
Na_2HPO_4	10 mM	11,7 g
adjust pH 7,2		

PBSMT

Skim milk	5 %
Triton X-100	0,1 %
dissolved in PBS	

This solution can be stored at -20 °C in a 50 ml falcon tube and thawed at 4 °C overnight.

PBT

Triton X-100	0,1 %
dissolved in PBS	

This solution can be stored at room temperature.

Antibody staining solution for IHC

BSA	0,1 %
dissolved in PBS and filter solution.	

This solution can be stored at -20 °C aliquoted in 2 ml epis.

AEC developing solution

1. 4 mg AEC dissolve in 500 µl dimethyl formamid (DMF)

2. Mix 2,1 ml 0,1 M Acetic acid with 7,9 ml 0,1 M sodium acetat. Finally reject 500 µl of this solution.

3. Mix the solutions prepared in step 1 and 2, finally filter solution and store at 4 °C in the dark.

4. Just before use give 5 µl H_2O_2.

Mowiol mounting media

Mowiol 4-88	20 g
PBS	80 ml;
NaN_3	0,02 %
glycerin (86%)	40 ml
adjust pH 7,5-8	

FACS staining buffer — 50 ml

FCS	2 %	1 ml
DNase I	0,25 µg/ml	1666 µl (7.5 µg/ml)
Prepare fresh with PBS.		

FACS buffer — 50 ml

FCS	2 %	1 ml
EDTA	2 mM	200 µl (0.5 M)
DNase I	0,25 µg/ml	1666 µl (7.5 µg/ml)
Prepare fresh with PBS.		

TNES — 100 ml

Tris pH 8,0	50 mM	5 ml (1 M)
EDTA pH 8,0	100 mM	20 ml (0,5 M)
NaCl	100 mM	2 ml (5 M)
SDS	1 %	5 ml (20 %)

10x TBE   5 l (10x)

Tris	0,1 M	605 g
H_3BO_3	0,1 M	310 g
EDTA	2 mM	37 g

add H_2O to 5000 ml

50x TAE   5 l (50x)

Tris base		1210 g
Glacial acetic acid		285,5 ml
EDTA		186 g

add H_2O to 5000 ml

2x Lucifarease assay buffer

Tricin	40 mM
$MgCO_3 \times Mg(OH)_2 \times 5\ H_2O$	2,14 mM
$MgSO_4$	5,4 mM
DMSO	0,2 mM

Heat solution to a max. 60°C.

1x Lucifarease assay buffer   200ml

2x Luciferase Assay Puffer		100 ml
DTT	1 M	6 ml
ATP	100 mM	1060 µl
CoA	54 mM	1 ml
Luciferin	25 mM	4 ml

Luciferase extraction buffer

$K-PO_4$ buffer (pH7,4)	100 mM
Triton X-100	1%
Glycerol	10%
EDTA	2 mM

add H_2O to 100 ml.

Store solution at room temperature, before use add to 10 ml of solution:

protease inhibitor (Complete Mini)   1 tablet

ELISA sample buffer

NaCl	150 mM
EDTA	66 mM
Tris/HCl pH7.4	10 mM
SDS	0.4 %
NP-40	1%

Store solution at room temperature, before use add to 10 ml of solution:

protease inhibitor (Complete Mini) 1 tablet

Dignam C — 100ml

Hepes (pH7,9)	20 mM	2 ml (1 M)
NaCl	42 mM	8,4 ml (5 M)
MgCl$_2$	1,5 mM	0,15 ml (1 M)
EDTA	0,2 mM	0,04 ml (0,5 M)
Glycerol	25 %	25 ml

Store solution at 4°C, before uses add to 10 ml of solution:

DTT	1 mM	10 µl (1 M)
PMSF	1 mM	200 µl (50 mM)
protease inhibitor (Complete Mini)		1 tablet

Complete solution can be stored at -20°C several months long.

RIPA buffer

Tris/HCl (pH 8)	50 mM
NaCl	150 mM
SDS	0,1 %
Deoxycholic acid	0,5 %
NP40	1 %

Store solution at room temperature, before use add to 10 ml of solution:

protease inhibitor (Complete Mini) 1 tablet

Bradford reagent   1 l

Coomassie Brillant Blue G-250	100 mg
Ethanol	50 ml
Phosphatic acid (85 %)	100 ml
add H_2O to 1000 ml	

4x Laemmli buffer   50 ml

Tris/HCl (pH 6,8)	0,125 M	12,5 ml (2 M)
SDS	4 %	10 g
Glycerin	20 %	13 ml
Bromphenolblue	0,004 %	1 spatula point

Store solution at room temperature, before use add
β-Mercaptoethanol 10 %

2x Laemmli-Urea sample buffer

Tris/HCl (pH 6,8)	62.5 mM
SDS	2 %
Glycerin	10 %
Bromphenolblue	0,00125 %
Urea	6 M

Dissolve in H_2O.
Store solution at -20°C, before use add
β-Mercaptoethanol 5 %

SDS-PAGE upper gel buffer 5 % (stacking gel)

Acrylamid/Methyl- Bisacrylamid (37,5:1)	2,5 ml (40 %)
Tris/HCl (pH 8,8)	5,1 ml (1,5 M)
SDS	100 µl (20%)
APS	400 µl (10%)
TEMED	27 µl l
H_2O	12,3 ml

SDS-PAGE lower gel buffer 10 % (separating gel)

Acrylamid/Methyl- Bisacrylamid (37,5:1)	11,25 ml (40 %)
Tris/HCl (pH 6,8)	11,25 ml (0,5 M)
SDS	450 µl (20%)
APS	400 µl (10%)
TEMED	30 µl
H_2O	21,75 ml

SDS-running buffer (10x)   1 l

Tris	25 mM	30,3 g
Glycin	192 mM	144 g
SDS	0.1 %	10 g

add H_2O to 1000 ml, adjust pH to 8,3.

Western Blot semi-dry transfer buffer   1 l

Tris	5,8 g
Glycin	2,9 g
SDS	1,85 ml (20 %)
Methanol	200 ml

Western Blot Stripping- Puffer   500 ml

Tris/HCl pH 6.7	31,25 ml (1 M)
SDS	50 ml (20 %)
β-Mercaptoethanol	3,45 ml

add H_2O to 500 ml

2.9.3 Histology chemicals and materials

Aceton	Roth, Germany
Acetic acid	Fluka, Switzerland
AEC (3-Amino-9-ethylcarbazol)	Sigma-Aldrich, Germany
Aquatex	Merck, Germany
BSA (Bovine Serum Albumin)	AppliChem, Germany
DAB (3,3`-Diaminobenzidine)	Sigma-Aldrich, Germany
DAPI (4',6-Diamidino-2-phenylindol)	Sigma-Aldrich, Germany
Ethanol	Sigma-Aldrich, Germany

Eosin (Sigma, MO)	Sigma-Aldrich, Germany
Eukitt quick-hardening mounting medium	Fluka, Switzerland
Glutaraldehyde	Sigma-Aldrich, Germany
Glycerol	AppliChem, Germany
Hematoxylin	Sigma-Aldrich, Germany
30% H_2O_2	J.T. Baker, Holland
Methanol	Merck, Germany
Methylbutan	Roth, Germany
Mowiol mounting media	self made
Sodium acetate	Sigma-Aldrich, Germany
$NiCl_2$	Sigma-Aldrich, Germany
Skim milk	AppliChem, Germany
Tissue-Tek OCT Compound	Sakura, Netherland
Triton X-100	Roth, Germany
Vectashield mounting media	Vector, CA
Xylol	Merck, Germany
Microscop slides Superfrost Plus	Menzel, Germany
Coverslip	Menzel, Germany
Cryomolds	Roth, Germany
Pap pen	KisKer GmbH, Germany

2.9.4 Special consumption items and equipment

Special materials and kits

Doxycycline (ICN Biomedicals, Inc.)
MMP-9 ELISA detection kit (MMPT90; R&D Systems, USA)
VEGF-A ELISA detection kit (MMV00; R&D Systems, USA)
High pure RNA isolation kit (Roche Diagnostics, Germany),
PicoPure RNA Isolation Kit (Arcturus Bioscience Inc., USA)
FastStart SYBR Green PCR Kit (Roche Diagnostics, Germany)
PE-conjugated monoclonal active caspase-3 antibody apoptosis kit (BD Biosciences, USA)
PE-conjugated Ki67 mouse a-human monoclonal antibody set (BD Biosciences, USA)
Trypan-blue (Sigma-Aldrich, Germany)
Proteaseinhibitor (Complete Mini, Roche Diagnostics, Germany)
Random hexamer primers (Invitrogen, USA)

RNase inhibitor (Invitrogen, USA)

Enzymes
DNase I (Quiagen, USA)
Proteinase K (AppliChem, Germany)
Collagenase/Dispase (Roche Diagnostics, Germany)
Taq DNA Polymerase (Invitrogen, USA)
MuMLV reverse transcriptase kit (Promega, USA)

Consumption items
Immobilon-P polyvinylidene fluoride (PVDF) membrane (Millipore, USA)
Chromatographic paper DE 81 (Whatman, USA)
Cell strainer (74 µm mesh, #3479 Costar, USA)

Electronic equipments
Inversionsmicroscope Leica DMIRB/E (Leica, Germany)
ApoTome fluorescence microscope, Axiovert 200M (Zeiss, Germany)
Stereomicroscope: MZ7.5 (Leica), Zeiss Stereo
Color camera ProgRes C14 (Jenoptik, Germany)
Cryostat CM1900 (Leica, Germany)
Lumat LB 9570 (Berthold Technologies, Germany)
LAS-1000 Luminescent Image Analyzer (FujiFilm, Japan)
Overgead shaker Reax 2 (Heidolph, Germany)
Photometer Ultraspec 3000 (Pharmacia Biotech, USA)
Semi-dry transfer cell / Tran Blot SD (Bio-Rad Laboratories, USA)
SpectraMax 190 microplate reader (Molecular Devices, USA)
FACSAria cell sorter instrument (BD Pharmingen, USA)
FACSCalibur flow cytometer (BD Pharmingen, USA)
Light cycler (Roche Applied Science, USA)
Software
AxioVision Rel.4.7 (Zeiss, Germany)
OpenLab, version 4.0.4 (Improvision, UK)
CellQuest (BD Pharmingen, USA)
ImageJ
Adobe Photoshop

2.9.5 Antibodies

Table 1. Antibodies

Name	Cat.no.	Company
IHC primary antibodies		
Fc-block	553142	BD Pharmingen
CD31/PECAM-1 (MEC13.3)	553370	BD Pharmingen
LYVE-1	103-PA50AG	ReliaTech GmbH
Laminin	L9393	Sigma-Aldrich
NG-2	AB5320	Millipore
Claudin-5	34-1600	Zymed
Occludin	71-1500	Zymed
Zonula occludens -1	61-7300	Zymed
Zonula occludens -2	71-1400	Zymed
α-SMA FITC (1A4)	F3777	Sigma-Aldrich
IHC secondary antibodies		
goat a-rat AF488	A-11006	Invitrogen
donkey a-rat AF-594	A-21209	Invitrogen
donkey a-goat AF488	A-11055	Invitrogen
goat a-rabbit TxRed	4050-07	Southern Biotechnology
goat a-rat HRP	474-1612	KPL
rat a-goat HRP	sc-2768	Santa Cruz
FACS antibodies		
Ki67-PE	556027	BD Pharmingen
cleaved caspase-3 PE	550914	BD Pharmingen
CD31/PECAM-1 FITC	558738	BD Pharmingen
CD105/Endoglin - biotin (MJ7/18)	13-1051-82	eBioscience
streptavidin-PE	12-4312-87	eBioscience
Western Blot		
Actin	A5060	Sigma
RelA	sc-372	Santa Cruz
c-Myc	sc-764	Santa Cruz
donkey a-rabbit HRP	711-036-152	Jackson ImmunoResearch

2.9.6 The list of primers used for RT-PCRs

Table 2. Primers sequences for RT-PCR:

gene	primer pair
β-Actin	forward 5'-GGTCAGAAGGACTCCTATGTG-3' reverse 5'-AGAGCAACATAGCACAGCTTC-3'
human-Myc	forward 5'-TTCCCCTACCCTCTCAACGACAG-3' reverse 5'-TCCTTACTTTTCCTTACGCACAA-3'
cytokeratin-8	forward 5'-CCT GCA GGC GGA GAT TGA AG-3' reverse 5'-TTG GTG GTG CGG CTG AAA GT-3'
Endoglin	forward 5'-TTT CCC GTC AGG CTC ACC ACC ACT-3' reverse 5'-GAG CAG GGC CCC AAT CAG GAA GG-3'
Pecam-1	forward 5'-AGG GGA CCA GCT GCA CAT TAG G-3' reverse 5'-AGG CCG CTT CTC TTG ACC ACT T-3'
β-catenin	forward 5'-GGC GTG CGC ATG GAG GAG ATA GTA-3' reverse 5'-GGG CAC CAA TGT CCA GTC AAA GAT-3'
Vimentin	forward 5'-ACA ATG ACG CCC TGC GCC AGG-3' reverse 5'-GCT CTC CTC GCC TTC CAG CAG C-3'
E-cadherin	forward 5'-CAG CTG CCC CGA AAA TGA AAA G-3' reverse 5'-GTA GGC GAT GGC AGC GTT GTA GGT-3'
VEGFR-2	forward 5'-GGG ACC ACA CTG CGC TCA CCT C-3' reverse 5'-GAC TCG GCC CTG GGA CAT CTC AC-3'
Tie-2	forward 5'-GTA CAA CGG CCA TTT CTC CTC ACT-3' reverse 5'-TAC ATT TCC CTC GCT TTT CTT CCT-3'

Table 3. Primers used for quantitative RT-PCR:

gene	primer pair
PBGD	forward 5'-GACCTGGTTGTTCACTCCCT-3' reverse 5'-TGGGTGAAAGACAACAGCAT-3'
TSP-1	forward 5'-GTGCTGCAGAATGTGAGGTT-3' reverse 5'-AAGAAGGACGTTGGTAGCTGA-3'
Ang-1	forward 5'-GCTAACAGGAGGTTGGTGGT-3' reverse 5'-GGTGGTGGAACGTAAGGAGT-3'

Ang-2	forward 5'-GTCAACAACTCGCTCCTTCA-3'	
	reverse 5'-GATTTCCGCACAGTCTCTGA-3'	
VEGFR-2	forward 5'-CCATTGGAGGAACCAGAAGT-3'	
	reverse 5'-CTCTTCTGATGCAAGGACCA-3'	
VEGF-A	forward 5'-CACTGGACCCTGGCTTTACT-3'	
	reverse 5'-TCACTTCATGGGACTTCTGC-3'	
Nrp-1	forward 5'-AGGACCATACAGGAGATGGC-3'	
	reverse 5'-AATAGACCACAGGGCTCACC-3'	
Dll4	forward 5'-GAGGTCCAAGCCGAACCTG-3'	
	reverse 5'-ATCGCTGATGTGCAGTTCACA-3'	
PDGF-β	forward 5'-GCACCGAAAGTTTAAGCACA-3'	
	reverse 5'-AAATAACCCTGCCCACACTC-3'.	

Quantitative real-time PCR primers for AM, CRLR, RAMP-2, RAMP-3 and eNOS were described elsewhere (Handa 2008; Ichikawa-Shindo 2008).

3. RESULTS

3.1. Model for conditional c-Myc expression in endothelial cells

In order to target c-Myc expression to the endothelium, we used a well-established inducible tetracycline-dependent transgenic system consisting of two components. One mouse strain is carrying a tetracycline responsive bi-directional minimal promoter–driven (tetO) human cDNA sequence of c-Myc encoding exons 2 and 3 and a luciferase reporter gene (tetO-Myc) (Marinkovic 2004). The second transgenic mouse line (Tie2-tTA) expresses the tetracycline-responsive transactivator under the control of the murine endothelial cell specific Tie2 promoter/enhancer elements (Figure 10.) (Deutsch 2008). In double transgenic animals (Tie2-tTA/tetO-Myc), tTA mediates the simultaneous transcription of both transgenes: the c-Myc proto-oncogene and the luciferase reporter gene. Presence of tetracycline or the synthetic derivative doxycycline prevents binding of tTA to the bidirectional promoter, thus resulting in the abrogation of c-Myc and luciferase transcription (tet-off system) (Gossen 1995).

In order to obtain Tie2-tTA/tetO-Myc double transgenic animals Tie2-tTA (C57BL/6 background), males were bred with tetO-Myc (NMRI outbred) females in the absence of doxycycline. No double transgenic animals were born from these crosses. Therefore, we concluded that c-Myc expression in endothelial cells during embryonic development results in embryonic lethality. However, Tie2-tTA/tetO-Myc double transgenic animals were born alive when pregnant mothers were kept under doxycycline.

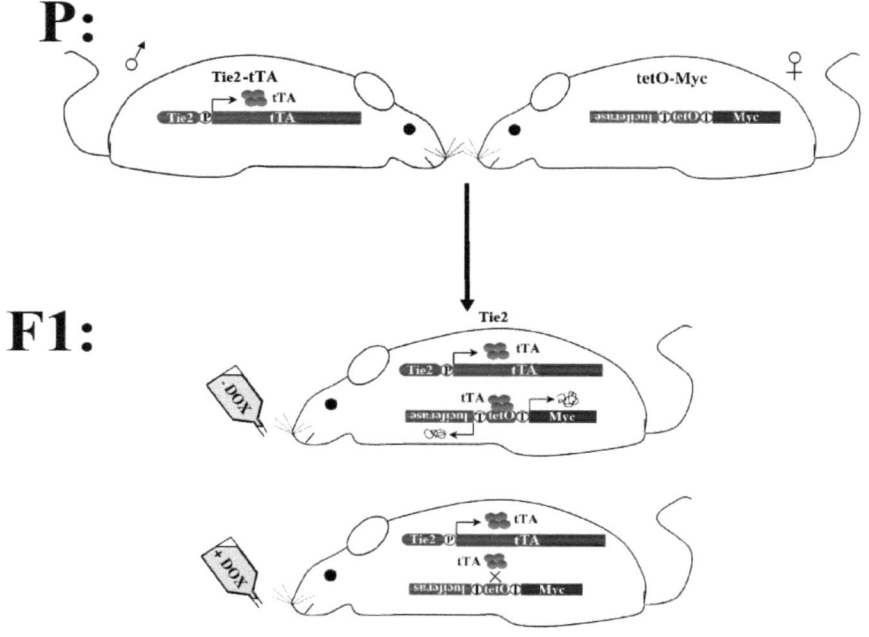

Figure 10. Conditional expression of transgenic c-Myc in endothelial cells using the tet-off system.

Two types of transgenic models were established. Tie2-tTA mice express the tetracycline-transactivator (tTA) under the control of the Tie2 promoter/enhancer (tet-off system). The tetO-Myc line contains constructs for the human c-Myc cDNA and a luciferase reporter gene under the control of the tetracycline responsive bi-directional promoter (tetO). In double transgenic animals (Tie2-tTA/tetO-Myc), tTA mediates the transcription of the c-Myc proto-oncogene and simultaneously the luciferase reporter gene. Upon doxycycline administration, tTA cannot bind to the bidirectional promoter, and expression of both, c-Myc and luciferase, is abrogated.

3.2. Endothelial cell-specific c-Myc expression in adult mice

3.2.1 Characterisation of the Tie2-tTA/tetO-Myc adult animals

No double transgenic animals were born from the crosses of Tie2-tTA males with tetO-Myc females in the absence of doxycycline. However, when pregnant mothers were kept under doxycycline during mating and pregnancy, Tie2-tTA/tetO-Myc double transgenic animals were born alive at a normal Mendelian frequency. Doxycycline was changed to water after birth on postnatal day 1 to induce transgene expression. With time, double transgenic mice developed tumors at different sites, for example in thymus, bulbourethral gland or at the nose.

3.2.2 Transgene expression in Tie2-tTA/tetO-Myc adult animals

Transgene expression was controlled for both at the RNA and protein levels in the tumors that developed in the thymi and the bulbourethral glands of the mice (Figure 11. A,B). The liver was not affected by the tumors and was therefore used as a mouse-specific negative control. Total RNA from organs was isolated, reverse-transcribed and human c-Myc-specific primers were used for the detection of the transgene. This analysis showed that human c-Myc mRNA was specifically detectable in the thymi and/or the bulbourethral glands of Tie2-tTA/tetO-Myc animals (#6537, #2998, #6540) (Figure 11. A). A single-transgenic animal (#2723) was used as a negative control. Extracts from a murine human c-Myc overexpressing lymphoma cell line served as a positive control for both experiments. Western blot analysis shows overexpression of the MYC protein in the thymi and the bulbourethral glands of Tie2-tTA/tetO-Myc animals as compared to the single transgenic control.

3.2.3 Survival of Tie2-tTA/tetO-Myc adult animals

When doxycycline was removed from the drinking water of neonatal Tie2-tTA/tetO-Myc double transgenic mice, these mice developed tumors in average with 22 weeks. The exact time point of death was somewhat variable between individual animals (Figure 12.). Double transgenic animals started to die around 17 weeks of age, and no double transgenic animal survived longer than 55 weeks.

This survival experiment was repeated two years later in an animal facility (N26O) with higher hygienic standards. The mice were transfered by *in vitro* fertilization to the N26O mouse facility. However, this time the experimental outcome was different. The time point of tumor induction was delayed and the tumor incidence was reduced yielding only 21% sick animals.

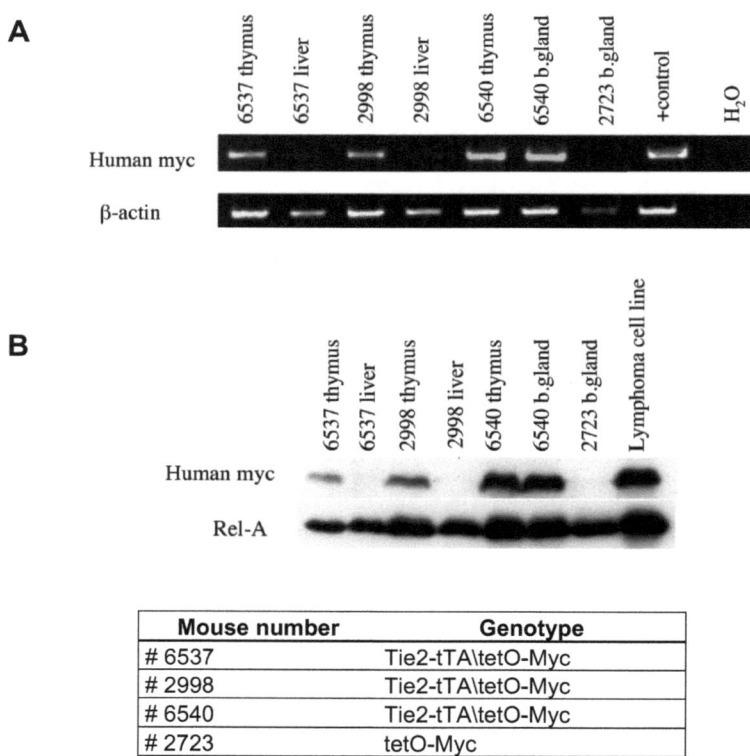

Figure 11. Transgene expression in Tie2-tTA/tetO-Myc animals.
(A) An RT-PCR analysis for the human c-Myc transgene is shown in upper panel. The lower panel shows a control RT-PCR for β-actin. Human c-Myc is only expressed in thymi and bulbourethral glands of Tie2-tTA/tetO-Myc animals. (B) Western blot analyses were performed using protein extracts from several double transgenic animals. The lower panel shows Rel-A as a loading control. Human c-Myc is highly expressed in thymi and bulbourethral glands of Tie2-tTA/tetO-Myc mice.

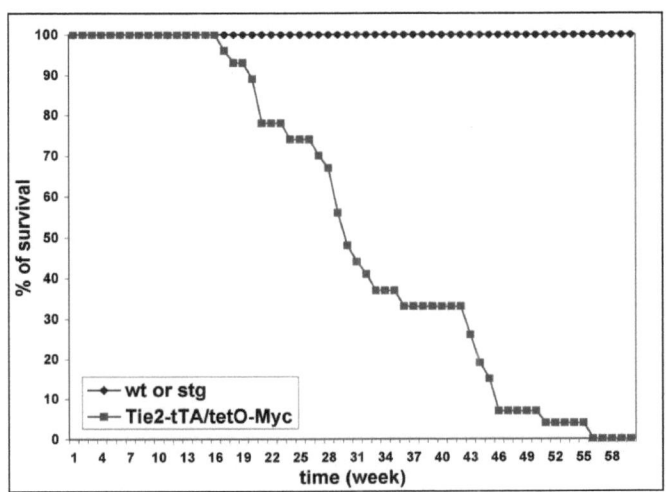

Figure 12. Survival of Tie2-tTA/tetO-Myc adult animals.
Kaplan-Meier plot shows the survival of Tie2-tTA/tetO-Myc mice. 37 double transgenic animals were analyzed. 20 wild type or single transgenic littermates were used in the control group. No double transgenic animals survived more than 55 weeks.

3.2.4 Gross pathology of Tie2-tTA/tetO-Myc animals

The gross anatomy of Tie2-tTA/tetO-Myc mice showed tumors in different organs of these animals. Most striking were tumors growing on the tail or on the legs of animals (Figure 13. A,C).

However, these animals appeared healthy aside from the growing tumor on their extremities. Other double transgenic animals exhibited excessive tumor growth in glandular organs like the thymus, the nose mucosa, the salivary gland and in the bulbourethral gland of male mice (Figure 13. B,C,D). These animals looked unhealthy, had ruffled fur and moved slowly. A few double transgenic animals showed even both types of tumors (Figure 13C).

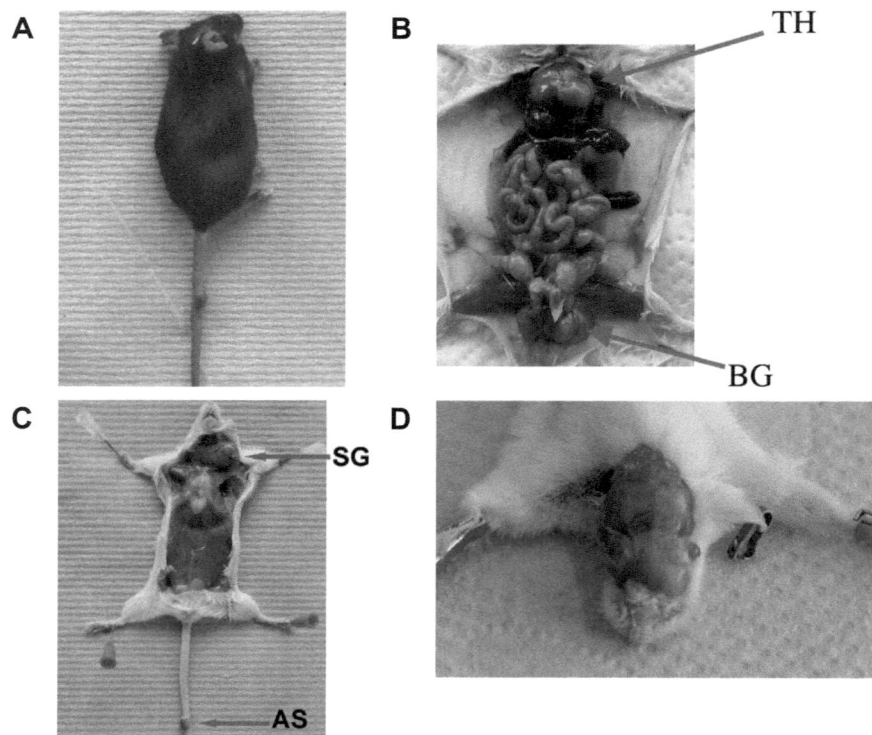

Figure 13. Gross anatomy of Tie2-tTA/tetO-Myc mice
(A) 30-week-old Tie2-tTA/tetO-Myc female mouse (#2726) with a tumor in the tail. (B) 21-week-old Tie2-tTA/tetO-Myc male mouse (#6540) showing a thymic (T) and a bulbourethral gland (BG) tumor. (C) 24-week-old Tie2-tTA/tetO-Myc double transgenic female mouse (#6786) shows tumors on the tail and in the salivary glands (SG). (D) 32-week-old double transgenic female mouse (#6552) with tumor of the mucous membrane. The tumor was extending from the nose mucous membrane infiltrating already into the eyes.

3.2.5 Tumor classification and statistics

The histological examination and survey of tumors were performed in cooperation with Dr. Ott, and Dr. Maier from the Animal facility, Ulm University.

Histological examination of the tumors growing on the extremities of animals revealed angiosarcomas (AS) (Figure 14A). Characteristic of angiosarcomas are the spindle-like tumor cells appearing in the tumor tissue (Figure 14C).

The majority of the tumor is filled with coagulated blood (Figure 14B,C). Angiosarcomas originate from endothelial cells of the vessels. In the case of blood vessels, the tumor is termed hemangiosarcoma and, in case of lymphatic vessels, lymphangiosarcoma.

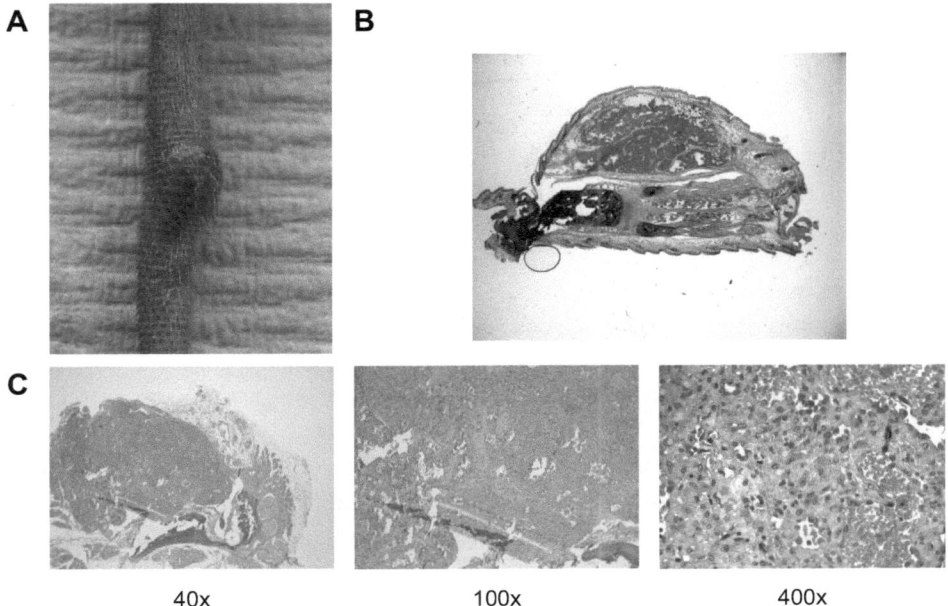

Figure 14. Angiosarcoma
(A) Photograph of a tail with angiosarcoma from a Tie2-tTA/tetO-Myc mouse. (B) HE-overview of the angiosarcoma. (C) HE-staining from an angiosarcoma on paraffin sections with different magnifications (40x, 100x and 400x). Tumor cells have a spindle-like shape. The tumor is filled with coagulated blood.

All tumors growing in glandular organs of double transgenic animals were found to be encapsulated. No metastases were detected in the liver, the lungs and the brain.

Hematoxilin-Eosin staining revealed adenomas, epithelial tumors of glandular and mucosal origin in the thymus, in the bulbourethral gland and in the nose mucosa. The bulbourethral gland from Tie2-tTA/tetO-Myc double transgenic animal showed a disorganized glandular structure with compact heterogenous cell aggregates with signs of necrosis and apoptosis (Figure 15B). Although metastases were not found in double transgenic mice, signs of malignancy were detected. It was found that tumor cells break into acini of the bulbourethral gland and into the bloodstream, too. Therefore this tumor is locally invasive, but does not form distant metastase.

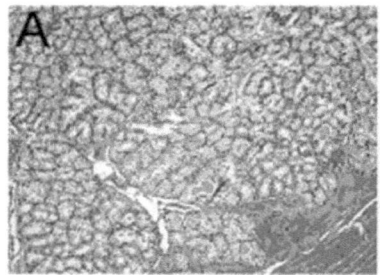 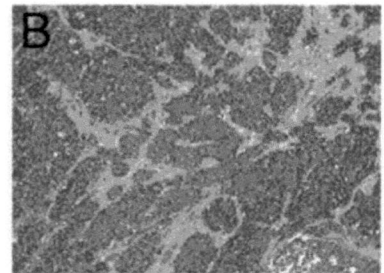

Figure 15. Adenoma
(A) HE-staining of the bulbourethral gland from a control mouse. The staining shows a complex healthy gland structure with serous and mucous cells. (B) HE-staining of the bulbourethral gland from a Tie2-tTA/tetO-Myc double transgenic mouse. The gland structure is still visible. However, it is disorganized with compact heterogenous cell aggregates with signs of necrosis and apoptosis.

Hematoxilin-Eosin staining of nose mucous membrane of a Tie2-tTA/tetO-Myc double transgenic mouse revealed a complex tumor, composed of glandular/adenomatous and spindle-like shaped cells (Figure 16B). Destructive tumor growth was observed as tumor cells were infiltrating other tissues. Bones were destroyed by the fast proliferating and infiltrating tumor cells (Figure 16A). Therefore this particular tumor is an adenocarcinoma.

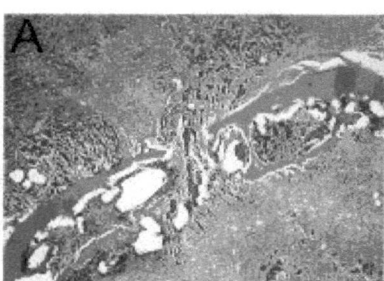 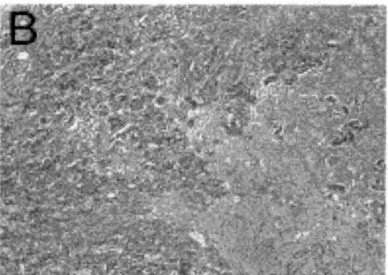

Figure 16. Adenocarcinoma of mucous membrane
(A) HE-staining of adenocarcinoma of the mucous membrane. The tumor is invading and destroying nose bones. Panel B shows the tumor mass with spindle-like and necrotic cells.

A total of 37 double transgenic animals were analyzed in the survival experiment (Table 4). Of these, 29 (77%) showed adenoma-like tumors, 3 (8%) both adenoma-like tumors and angiosarcomas, and 5 (15%) angiosarcomas only. No double transgenic animal survived more than 55 weeks.

Tumor type	Percentage
Angiosarcomas	15%
Adenoma-like tumors	77%
Angiosarcomas + Adenomas	8%

Table 4. Statistics of tumor development

Statistics was designed from a total of 37 Tie2-tTA/tetO-Myc mice.

3.2.6 Tumor regression upon transgene inactivation

Several previous reports suggested that the inactivation of key players in the process of tumorigenesis might result in tumor regression (Felsher 1999a; Marinkovic 2004). Therefore we also tested the idea whether suppression of transgenic c-Myc expression in Tie2-tTA/tetO-Myc mice *in vivo* would result in a reversal of the tumor phenotype.

Angiosarcoma-bearing mouse with advanced tumor size were treated with doxycycline in the drinking water. After 4 weeks, doxycycline treatment already reduced the tumor size by at least 50% (Figure 17A,B). Longer doxycycline treatment, however, did not result in a further regression of the tumor. Doxycycline treatment of adenoma-bearing mice resulted similar: upon 4 weeks of doxycycline-treatment, tumor size was reduced by at least 50%.

These results show that angiosarcoma and adenoma development is partially reversible in double transgenic mice. Turning off the transgene expression resulted only partially in tumor regression. Thus, tumor development is only in part dependent on the c-Myc transgene expression. In order to obtain a complete regression of the tumor, additional factors are required.

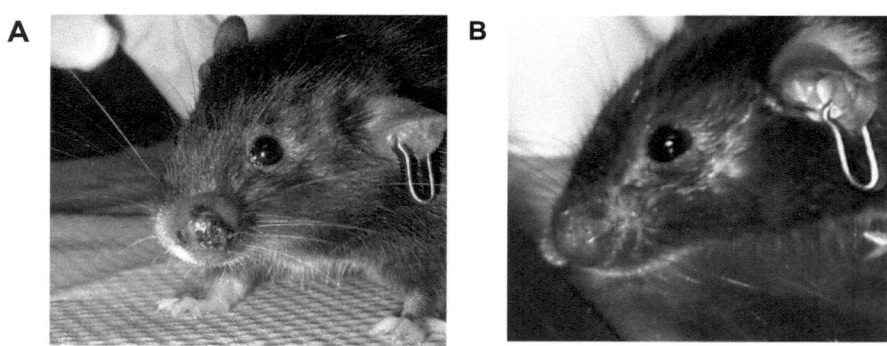

Figure 17. Tumor regression upon c-Myc inactivation

(A) A 23-week-old double transgenic male mouse (#8273) shows almost bursting angiosarcoma on the mouth. Mouse was treated with 2mg/ml doxycycline for 4 weeks. (B) The same mouse after 4 weeks of doxycycline treatment with a partially regressed angiosarcoma.

3.2.7 Establishment of tumor cell lines

Establishment of cell lines from tumors opens the possibility to analyse oncogenic principles *in vitro*. Therefore, our aim was to establish cell lines from the mouse angiosarcomas and adenomas. This work was performed in collaboration with Dr. Silke Brüderlein from the Department of Pathology, Ulm University.

Both tumor types were used for cell line isolation (Table 5). The different cell populations growing *in vitro* after isolation were analyzed for human c-Myc expression by RT-PCR (Figure 18). However, human c-Myc transgene expression could be detected only in a dendritic cell-like cell subpopulation of one tumor sample (09/05), derived from an angiosarcoma. Therefore this subpopulation was further enriched in culture and used for subsequent experiments.

Tumor type	Angiosarcoma	Adenoma
Origin	nose mucosa	bulbourethral gland
Mouse no.	# 8273 M	# 8271 M
Cell line no.	09/05	08/05

Table 5. Cell line isolation
Cell lines were isolated from an angiosarcoma and an adenoma.

3.2.8 Characterization of the tumor cell line

In order to identify the origin of the established cell line (09/05-dendritic), the expression of endothelial, epithelial and mesenchymal markers were analysed by RT-PCR (Figure 19).

I used bEnd.5, EpRas and XT cells as control cell lines for RT-PCR. The bEnd.5 mouse endothelioma cell line was used as a control for endothelial marker expression. EpRas cells are mouse epithelial cells, and XT cells are mouse cells with more mesenchymal characteristics.

The following cell-type-specific markers were used: Tie2, Pecam-1, VEGFR-2 and Endoglin are endothelial cell-specific markers, Vimentin is expressed in mesenchymal cells, E-cadherin and Cytokeratin-8 are markers of epithelial cells.

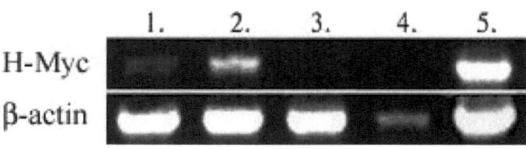

samples	cell line	subpopulations
1.	08\05	undifferentiated cells
2.	09\05	big, round shaped, dendritic cell-like cells
3.	09\05	fibroblasts + a few big round shaped cells
4.	09\05	epithelial cells
5.	lymphoma	mouse lymphoma cell line, positive control

Figure 18. Tumor cell line isolation
Analysis of human c-Myc expression. RT-PCR shows human Myc expression only in the positive control and in dendritic cell-like cell subpopulation obtained from the cell line 09/05. The lower panel shows the control RT-PCR for β-actin.

The cell line 09/05-dendritic expresses both endothelial cell markers, VEGFR-2 and Endoglin. In contrast, the other endothelial cell-specific markers, Pecam-1 and Tie2 are not expressed by the cell line 09/05-dendritic. Vimentin is expressed by a wide variety of mesenchymal cell types. In addition to XT, Vimentin expression was also found in the bEnd.5 endothelial cell line as well as in the 09/05-dendritic line. However, the cell line 09/05-dendritic did not express the epithelial markers E-cadherin and cytokeratin-8.
Therefore the cell line 09/05-dendritic does not show an epithelial-like phenotype, but rather exhibits a mesenchymal origin, with some typical endothelial cell markers.

After several weeks of propagation, the cell line 09/05-dendritic had lost its luciferase expression (Figure 20). However, this result is supported by the observations of Urban Deutsch. He found that in isolated endothelial cells from mouse using the Tie2 promoter to express a transgene, transgene expression is lost *in vitro* (unpublished data). Therefore we were interested, whether we could regain transgene expression under *in vivo* conditions.

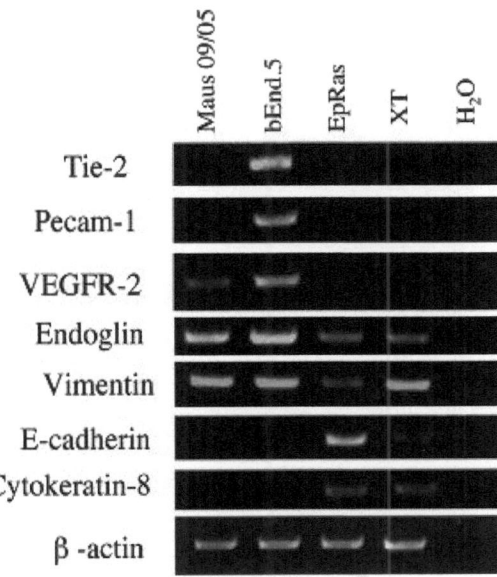

Figure 19. Characterization of the cell line 09/05-dendritic
The expression of cell-type-specific markers in the cell line 09/05-dendritic was assessed by RT-PCR. The bEnd.5 endothelial, EpRas epithelial, and XT mesenchymal cell were used as control mouse cell lines. The lowest panel shows a control RT-PCR for β-actin.

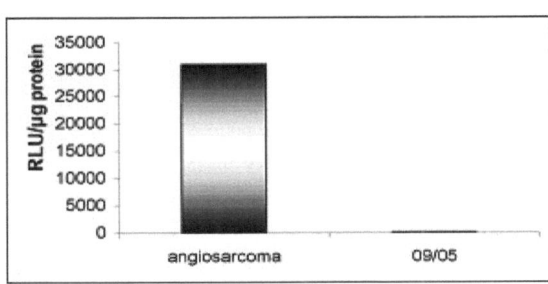

Figure 20. Luciferase activity of the cell line 09/05-dendritic
Luciferase activity was measured in the original angiosarcoma and in the isolated c-Myc expressing cell line 09/05-dendritic. During several weeks of propagation the cell line 09/05-dendritic had lost its luciferase expression.

In our experimental setup, Rag-2 -/- immunodeficient mice were transplanted with the cell line 09/05-dendritic. Rag-2 -/- mice fail to produce mature B or T lymphocytes, however they still have dendritic cells, neutrophils and macrophages. The transplanted animals were divided into two groups. The control group was not treated with doxycycline, a condition when the transgenic system could be active in the transplanted cells. Tumor growth could occur in these animals. The other group of animals was treated with doxycycline in the drinking water for 6 weeks. Thus the tet-off system is inactive and we did not expect tumor growth. After 6 weeks the mice were sacrificed, tumors were harvested and luciferase activity was measured.

We found tumors in all mice regardless of whether they were treated with doxycycline or not (Figure 21). Given that luciferase expression was not reactivated in the tumors *in vivo* (data not shown), we assume that also expression of c-Myc was independent of the tTA-system. The cell line therefore has lost its regulable system of gene expression.

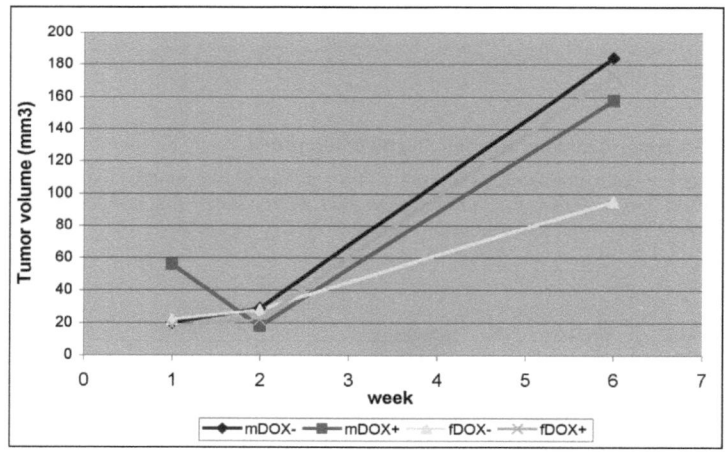

Figure 21. Tumor growth in Rag-2 -/- mice
Rag-2 -/- mice were injected subcutaneously with the cell line 09/05-dendritic. Mice were not treated with doxycycline; male DOX- (n=5, dark blue line) and female DOX- (n=4, yellow line). Mice in the other group were treated with doxycycline 6 weeks long; male DOX+ (n=5, pink line) and female DOX+ (n=4, light blue line). No differences were found in tumor growth between the two groups of mice.

3.2.9 Subsummary I

Since Tie2-tTA/tetO-Myc mice died during embryogenesis we made use of the employed regulatory feature, the tet-off system. When embryonic development took place under doxycycline administration, double transgenic mice were born and lived in average for 36 weeks. Starting from weeks 22, these mice began to develop tumors in the thymus, the bulbourethral gland, and at the nose. According to histological analyses, these tumors were angiosarcomas, adenoma-like tumors, rarely, adenocarcinomas.

Partial tumor regression was observed after the application of doxycycline for 4 weeks. Longer application of doxycycline did not result in a further decrease of tumor mass, indicating some loss of oncogene-addiction. Indeed other studies on c-Myc induced hepatocellular cancer and osteosarcoma showed a fully tumor regression upon short or long time c-Myc inactivation (Jain 2002; Shachaf 2004).

In a collaborative effort, cell lines from these tumors were established and further characterized. Of 4 different cell lines from 2 animals, only one showed luciferase expression and was further propagated. This cell line expressed the endothelial markers VEGFR-2 and Endoglin as well as the mesenchymal marker Vimentin. However, luciferase expression in this cell line was no longer regulatable by doxycycline application *in vitro* and *in vivo*, i.e. xenografts developed independent of doxycycline administration.

Since the spontaneous tumor formation was not reproducible in a second survival experiment, we had to conclude that the transgene manipulation, required for changing of the animal facility, may have influenced the outcome of this experiment, additional to the changing environmental conditions.

We therefore focused on the search for the embryonic lethality of c-Myc when driven by the Tie2 promoter.

3.3. Characterisation of Tie2-tTA/tetO-Myc embryos

3.3.1 Transgene expression in Tie2-tTA/tetO-Myc embryos

F1 embryos arising in crosses of Tie2-tTA males with tetO-Myc females were used for experimental investigations.

Human transgenic c-Myc and reporter gene expression was assessed in Tie2-tTA/tetO-Myc embryos. Transgene activity was first demonstrated by monitoring the luciferase reporter gene activity. Luciferase expression in E14.5 embryo sections was visualized using luciferin-containing buffer in a chemiluminescence detection system. Luciferase activity was observed throughout the embryonic body, and the highest activity was found in the lungs (Figure 22A). Therefore, further transgene expression analyses were performed using embryonic lung tissue samples.

In order to control for proper functioning of the tet-off system, during pregnancy, mice were treated either with doxycycline via the drinking water, or left untreated. Luciferase expression from E14.5 embryonic lung protein extracts was assessed, and high luciferase expression levels were observed exclusively in non-treated c-Myc expressing embryos (Figure 22B).

The tTA-mediated c-Myc expression was analysed in total RNA isolated from the lungs of E14.5 embryos. Human c-Myc specific primers were used for RT-PCR analysis. This analysis revealed that c-Myc expression was specifically detectable in Tie2-tTA/tetO-Myc embryos from mothers not treated with doxycycline (Figure 23A).

Finally, using Western blot analyse it was possible to detect slightly increased c-Myc protein levels in double transgenic embryos as compared to wild type or single transgenic controls (Figure 23B).

These results demonstrate that the human c-Myc transgene is specifically expressed in Tie2-tTA/tetO-Myc embryos and the tet-off system can be regulated by doxycycline.

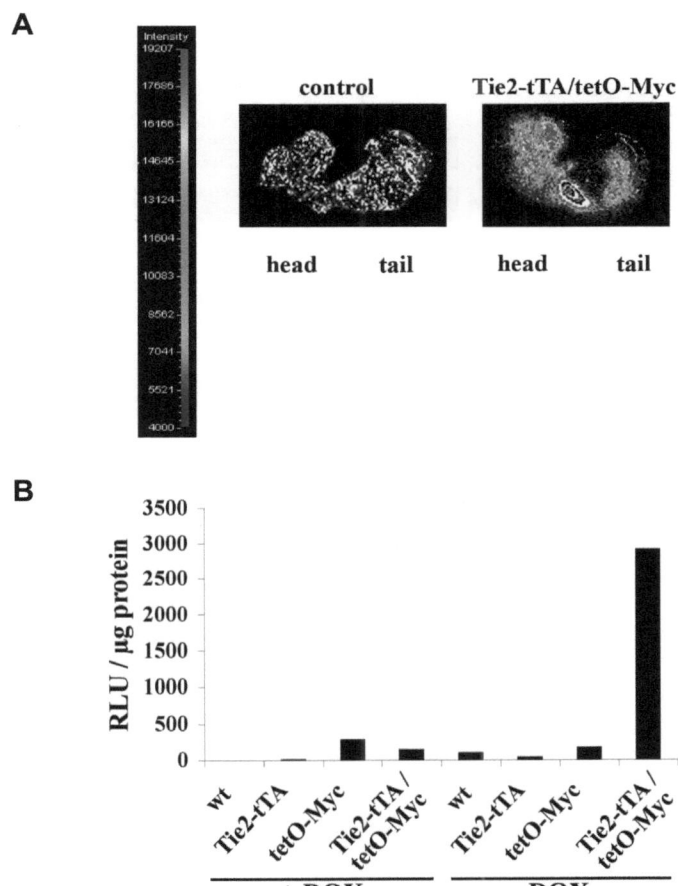

Figure 22. Reporter gene expression and control of the tet-off system in Tie2-tTA/tetO-Myc embryos

(A) Luciferase reporter gene expression on E14.5 embryonic tissue sections was visualized using a chemiluminescence detection system. The highest luciferase expression was observed in the lungs of Tie2-tTA/tetO-Myc embryos. Representative control and double transgenic animals are depicted. (B) Luciferase activities were measured in E14.5 embryonic lung protein extracts. The respective genotypes and doxycyline treatment are indicated. A representative experiment is shown.

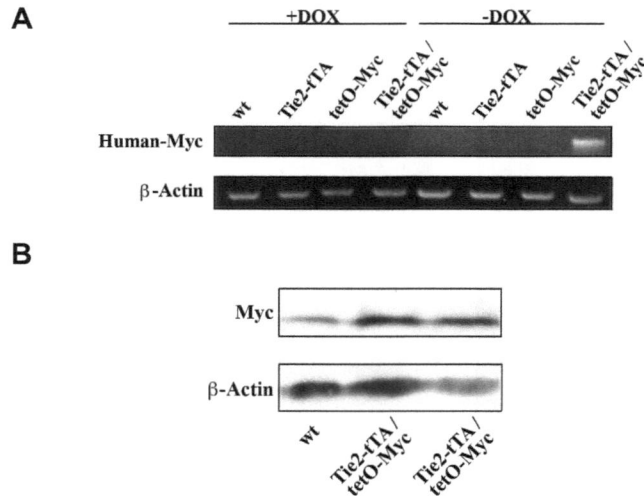

Figure 23. Transgene expression and control of the tet-off system in Tie2-tTA/tetO-Myc embryos

(A) Expression of the human-c-Myc transgene was assessed in the lungs of E14.5 day embryos by RT-PCR. The lower panel shows a control RT-PCR for β-actin. The genotype of the mouse embryos is indicated on top. Pregnant mothers were kept with or without doxycycline to control for the function of the tet-off system. RT-PCR shows expression of human c-Myc only in the non doxycycline-treated Tie2-tTA/tetO-Myc embryo. (B) Western blot analysis of c-Myc protein expression in E14.5 embryonic lungs of mice with the indicated genotypes. Expression of actin was determined as a control.

3.3.2 Endothelial cell-specific c-Myc expression causes embryonic lethality

The analysis of offspring from mothers not treated with doxycycline during pregnancy revealed that no double transgenic animals were born. However, in the absence of doxycycline, Tie2-tTA and tetO-Myc single transgenic and wild type animals were born similar to expected mendelian frequency. When the pregnant mothers were treated with doxycycline in the drinking water, Tie2-tTA/tetO-Myc mice were born at mendelian frequency (Figure 24A).

In order to determine more precisely when double transgenic embryos die, analyses of embryos were performed at different time points of embryogenesis. Embryos were classified as alive or dead by means of the beating heart. The genotype was determined both by luciferase measurements as well as PCR analyses. Luciferase measurement was performed from embryonic tissue protein extracts. Transgenic animals overexpressing c-Myc in endothelial cells were present at normal frequency and with a normal appearance up to E12.5 of gestation. The exact time point of death was somewhat variable between individual embryos. Some of the Tie2-tTA/tetO-Myc double transgenic embryos had already died at E14.5, most of the animals were dead at E16.5, and not a single living double transgenic animal had survived until E18.5 (Figure 24B).

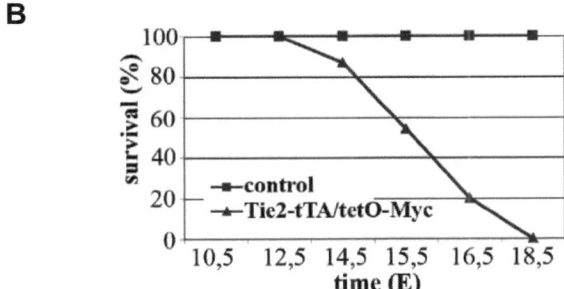

A

	-DOX	+DOX
wt	13	14
tetO-Myc	21	17
Tie2-tTA	11	13
Tie2-tTA/tetO-Myc	0	17
Total	45	61

Figure 24. Endothelial cell specific c-Myc expression results in embryonic lethality
(A) Genotypes of progeny from the crosses between Tie2-tTA and tetO-Myc mice in the presence or absence of doxycycline. (B) Survival rate of Tie2-tTA/tetO-Myc transgenic embryos derived from crosses without doxycycline. A total of 66 double transgenic Tie2-tTA/tetO-Myc embryos were scored out of 231 embryos in total at the indicated days of embryonic development. Single transgenic or wild type animals were used as a control. 50% of the double transgenic embryos survive till E15.5, no double transgenic embryos survive longer than E18.5.

Gross pathological examination showed the occurrence of a widespread subcutaneous edema on all Tie2-tTA/tetO-Myc embryos (Figure 25A, upper panel, embryo) and hemorrhages (Figure 25A, lower panels). The size of Tie2-tTA/tetO-Myc embryos was normal as compared to wild type or single transgenic littermates. Histological analyses of double transgenic animals by Hematoxylin and Eosin staining further demonstrated the subcutaneous edema (Figure 25B).

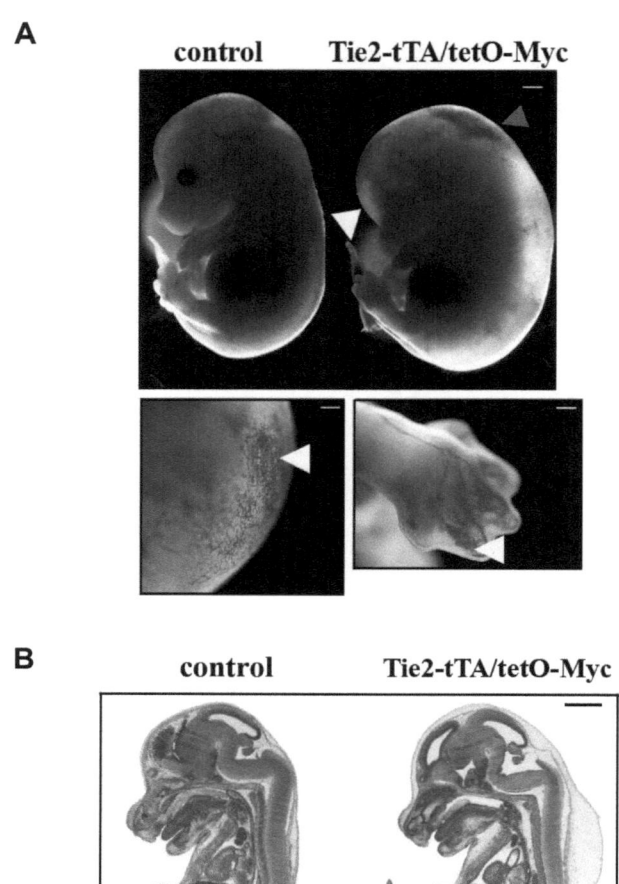

Figure 25. Gross examination of whole embryos

(A) Gross anatomical analysis of whole embryos at E16.5. Widespread subcutaneous edema (red arrowhead, upper panel) and multifocal hemorrhages (white arrowheads) on the extremities (upper right and lower right panels) and head (lower left panel) are evident. None of these alterations was seen in control littermates (embryo on the left upper panel). Scale bars 2 mm in upper panel, 400 µm in lower left and 350 µm in lower right panel. (B) HE staining of sagittal sections of E15.5 embryos. Tie2-tTA/tetO-Myc double transgenic embryos exhibit widespread subcutaneous edema. Scale bar 2 mm.

3.3.3 Possibly origins of vascular permeability

Given the severe edema phenotype of the Tie2-tTA/tetO-Myc double transgenic embryos, the question arises which defect induces the formation of the edema?

One possible reason of developing edema could be an immature embryonic dermal vasculature. During vessel maturation, associated support cells become recruited along developing vessels like pericytes and SMCs. Also the vascular basement membrane is formed during this phase of vascular development.

Edema could occur when during one of these processes failures happen. The developing immature vasculature would have higher permeability, or would be leaky.

In order to test this hypothesis, we analyzed the morphology of vessels.

The composition of the vessel with endothelial cell layer, basement membrane, vascular mural cell cover, I define in my thesis as vessel morphology.

Double immunofluorescence stainings were performed with a pericyte marker, the NG2-specific antibody. Endothelial cells and blood vessels were stained with the anti-CD31/PECAM-1 antibody. This analysis did not reveal any obvious differences between subcutaneous tissues of control mice and subcutaneous edematous tissues of double transgenic embryos (Figure 26A).

In addition, the organization of smooth muscle cells covering dermal arteries was analyzed with an α-SMA antibody. Microvessels are not covered with α-SMA positive vascular mural cells. Again, no differences were observed between control and double transgenic animals (Figure 26B).

The integrity of the basement membrane in dermal vasculature was also investigated. Thorough examination of laminin expression and organization in control and double transgenic animals did not reveal any apparent differences (Figure 26C).

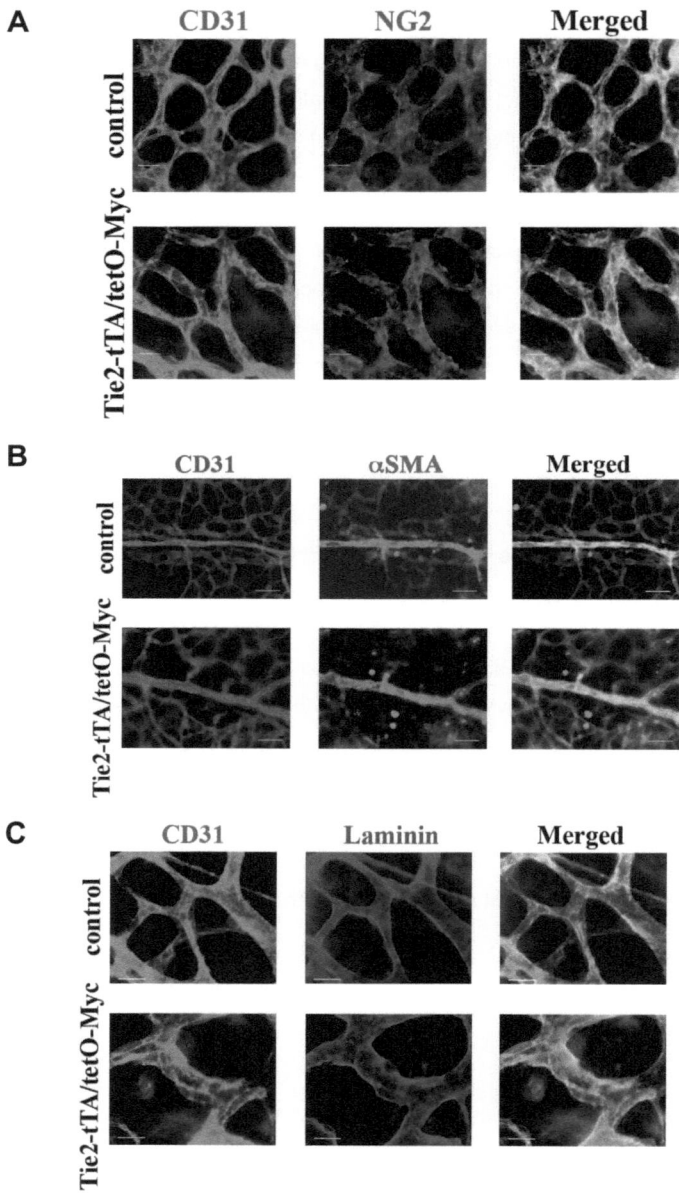

Figure 26. Characterisation of blood vessel structure in embryonic skin

(A) Pericyte coverage of embryonic blood vessels was visualized in E14.5 embryos by whole-mount immunofluorescence. Endothelial cells were stained with a CD31/PECAM-1 antibody (green), pericytes were stained with a NG2 antibody (red). Scale bar 20 µm. (B) SMC cover examination on E14.5 embryonic back skin tissue by whole-mount immunofluorescence. Double labeling for CD31/PECAM-1 (red) and α-SMA (green) for the indicated genotypes is shown. Scale bar 63 µm. (C) Basement membrane analysis on E14.5 embryonic back skin by whole-mount immunofluorescence. Double labeling for CD31/PECAM-1 (green) and Laminin (red) for the indicated genotypes is shown. Scale bar 20 µm.

Matrix metalloproteinases (MMPs) are responsible for remodeling of the extracellular matrix during development and disease (Rundhaug 2005). Since it was shown that c-Myc induces MMP-9 expression in endothelial cells, MMP-9 protein levels in Tie2-tTA/tetO-Myc double transgenic and wild type embryos were assessed (Magid 2003). A slight increase of MMP-9 protein expression levels in the lungs of double transgenic Tie2-tTA/tetO-Myc embryos at E15.5 was noticed (Figure 27).

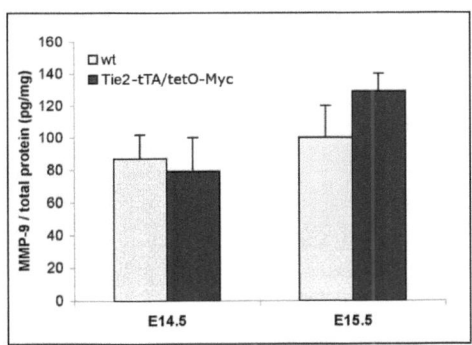

Figure 27. Matrix metalloproteinase-9 protein expression by embryonic lungs
Total MMP-9 protein expression was determined by ELISA measurement in embryonic lung extracts. The MMP-9 concentration was normalized to the total protein in the tissue samples. Double transgenic embryos show a slightly elevated expression of MMP-9 as compared to wild type embryos at E15.5. Error bars represent SEM.

The analysis shown so far indicate that Tie2-tTA/tetO-Myc double transgenic embryos at E14.5 have normal vessel morphology with fully developed basement membranes and an intact pericyte and smooth muscle cell cover. Matrix metalloproteinase-9 expression and extracellular matrix remodeling is not significantly elevated in double transgenic embryos.
An additional possible reason for the increased vascular permeability could be an impaired tight junction formation between endothelial cells. Therefore, the expression of different junctional proteins like ZO-1, ZO-2, claudin-5 and occludin (red) was assessed by

immunofluorescence stainings in embryonic dermal vasculature (Figure 28). Blood vessels were stained for the CD31/PECAM-1 (green) endothelial cell marker. Nuclei were counterstained with DAPI (blue). These analyses revealed no significant changes in the expression of the tested tight junction proteins.

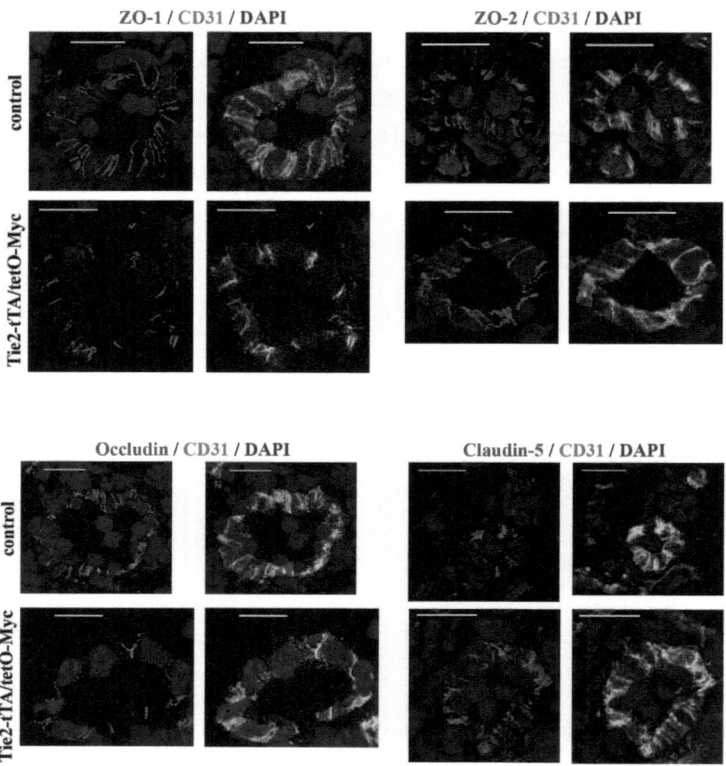

Figure 28. Analysis of junctional connections between endothelial cells of dermal blood vessels at E14.5

Immunofluorescence stainings show tight junctional protein expression in embryonic dermal blood vessels for the indicated genotypes. Endothelial cells were stained with CD31/PECAM-1 (green), ZO-1, ZO-2, occludin, claudin-5 junctional proteins (red) were stained, and cell nuclei were visualised with DAPI (blue). Scale bars 20 μm.

3.3.4 Analysis of the lymphatic vessels

Since the blood vasculature of Tie2-tTA/tetO-Myc double transgenic embryos developed properly, it was suspected that an increased vascular permeability could arise due to lymphatic dysfunction. Therefore the lymphatic vasculature of E14.5 embryos was examined with a LYVE-1-specific antibody, which specifically stains lymphatic vessels (Figure 29A).

This analysis revealed no differences between lymphatic vessels of control and Tie2-tTA/tetO-Myc embryos with respect to vessel architecture, complexity and sprouting. The assembly of the vascular tree I define as vessel architecture. Further immunohistological analyses showed that lymphatic vessels were intact, and no infiltrating blood cells appeared in lymphatic vessels adjacent to blood vessels (Figure 29B). The appearance of lymphatic vessels was quantitatively analyzed on double-immunofluorescence stained embryonic tissue sections. These examinations revealed that Tie2-tTA/tetO-Myc embryos showed unaltered numbers of normally sized lymphatic vessels in the dermis and at the jugular area (Figure 30A).

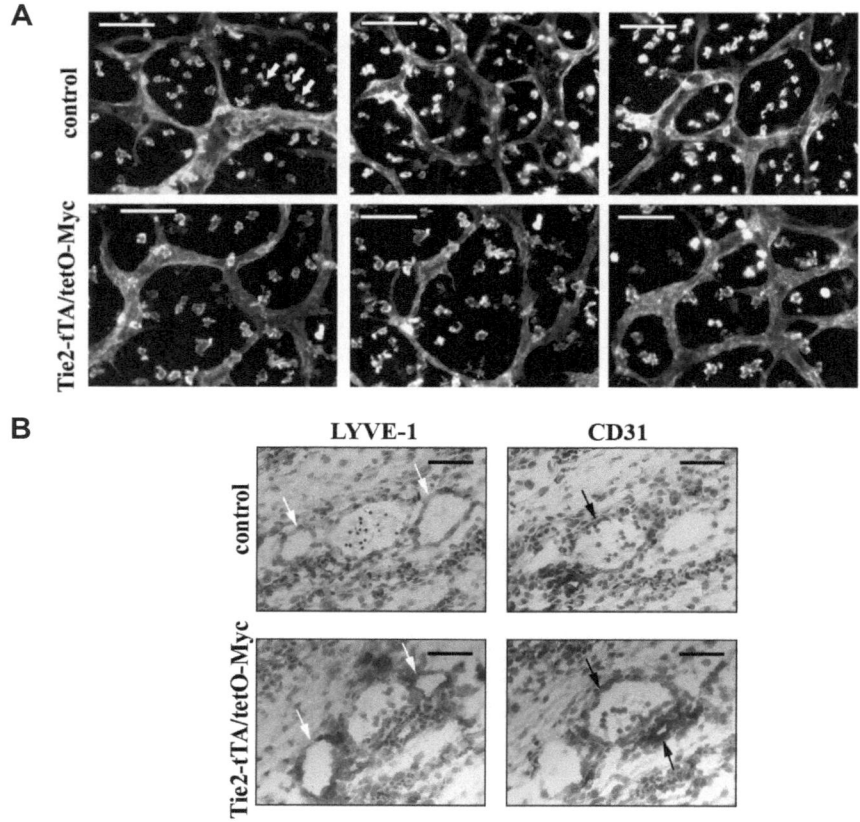

Figure 29. Immunohistochemical analysis of lymphatic vessels.

(A) The lymphatic vasculature was visualized with LYVE-1 specific antibodies by whole-mount immunofluorescence stainings of embryonic back skin of E14.5 animals. Three different images of the skin lymphatic vessels from the indicated genotypes are shown. In addition to lymphatic endothelial cells, also macrophages stain positive for LYVE-1 (arrows). Scale bar 100μm. (B) Left panels show LYVE-1 staining of lymphatic vasculature in E14.5 embryos. Lymphatic vessels are indicated by white arrows. Right panels show CD31/PECAM-1 staining of blood vessels from adjacent sections for the indicated genotypes. Blood vessels are assigned by black arrows. Scale bar 160 μm.

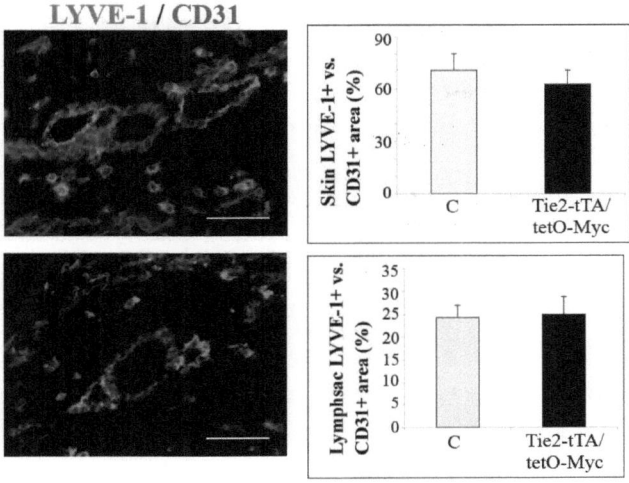

Figure 30. Quantitative analysis of lymphatic vasculature
Left panels show immunofluorescence stainings performed with LYVE-1 (green) lymphatic and CD31/PECAM-1 (red) blood vessel specific antibodies on E14.5 embryonic sections. Scale bar 100 µm. The quantification of LYVE-1 and CD31/PECAM-1 staining is shown in the right graphs. Staining quantification was performed from embryonic dermal lymphatic vasculature and from the jugular area, n=6 for the indicated genotyps.

3.3.5 Analyses of dermal blood vessel architecture

Due to the detailed analysis of vessel morphology we could notice interesting alterations in the assembly of the dermal vasculature.

To get more details on the alterations of the developing vessel architectur further analyses were performed at different embryonal ages. The development of the primary capillary plexus was analysed by E9.5 embryos. Later vascular remodeling was evaluated by E13.5 and E14.5 embryos.

To our analyses we used the technique of whole mount staining. Whole embryos were stained using antibodies specific for the endothelial cell marker CD31/PECAM–1.

Analysis of E9.5 embryonic vasculature of control and double transgenic embryos showed no obvious differences, suggesting that vasculogenesis occurred normally until this age of development (Figure 31A). However, the vascular network in E14.5

c-Myc overexpressing embryos was significantly different (Figure 31B). Tie2-tTA/tetO-Myc embryonic dermal vasculature formed fewer branches, indicating defective angiogenesis or enhanced pruning of vessels (Figure 31B).

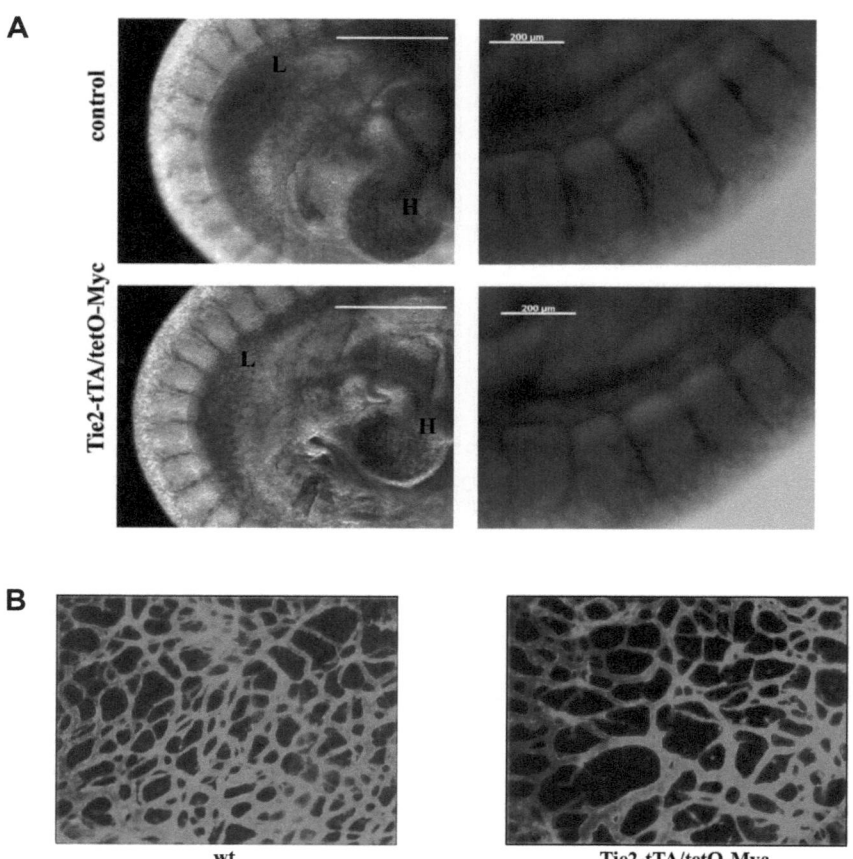

Figure 31. Whole-mount analysis of embryonic vasculature

(A) The vascular system of E9.5 embryos was visualized by whole-mount staining for CD31/PECAM-1. Representative images were taken from middle of the embryo (left), and a higher magnification from somites is shown at the right side. Scale bar 500µm (left) and 200µm (right). L: hind limb, H: heart (B) Representative images of CD31/PECAM-1 whole-mount fluorescence immunohistochemistry stainings used for morphometric analyses. Identical magnification was used for both pictures. In all cases, identical areas at the back of the embryos were analyzed. The obvious morphological differences between control and Tie2-tTA/tetO-Myc embryos were quantified. Scale bars 100 µm.

The obvious architectural alterations in the dermal vasculature of E14.5 double transgenic embryos were statistically analyzed in collaboration with the Institute of Stochastics of the Ulm University (director Prof. Volker Schmidt). The analyses were performed on distinct parameters that define the vascular network (Figure 32). The degree of branching was determined as the number of branching points per area as well as the number and mean lengths of vessels. These analyses revealed a striking reduction in branching points and correspondingly fewer vessels in the double transgenic animals (Figure 32A,B). Consistent with this sparsity of segments, the mean length of vessel in-between branching points was increased (Figure 32C).

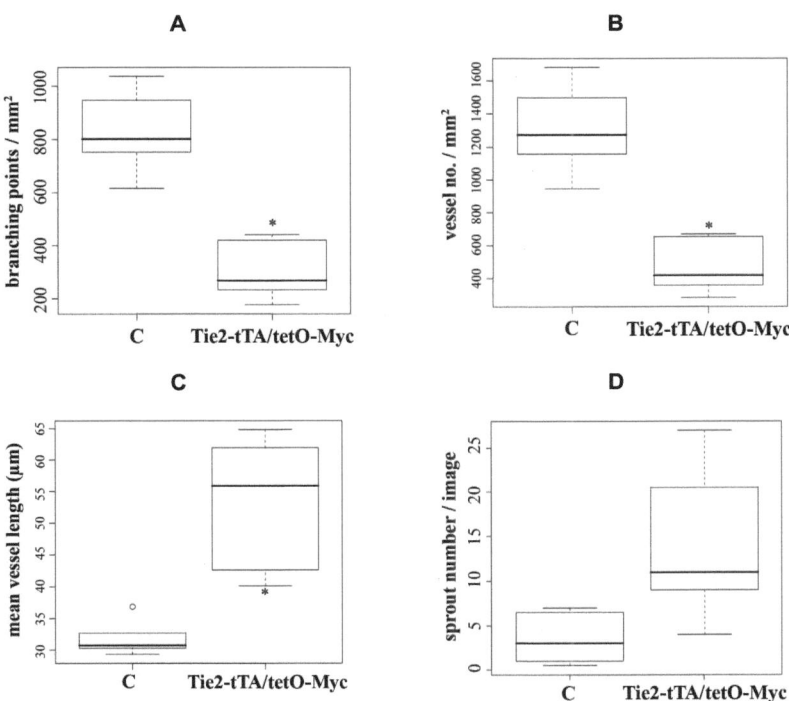

Figure 32. Morphometric analyses of back skin blood vessels from control and Tie2-tTA/tetO-Myc embryos at E14.5

Branching points per area, vessel number per area, mean vessel length and number of sprouts per image were quantified by control (wt or single transgenic) and Tie2-tTA/tetO-Myc animals. In both groups a total of n=6 animals and two images per animal were analyzed. *P<0.01 Wilcoxon test.

In addition, we also counted vessel sprouts that had not successfully connected to another vessel. The number of such sprouts was clearly increased in Tie2-tTA/tetO-Myc animals as compared to controls (Figure 32D) indicating an increase in ongoing but not yet completed angiogenic activity.

However, statistical analysis of E13.5 embryonic back skin vascular architecture revealed no significant differences in the number of branching points, number of vessels and the mean length of vessels between the control and double transgenic embryos (Figure 33). These data suggest that blood vessels are formed normally up to E13.5 in Tie2-tTA/tetO-Myc embryos.

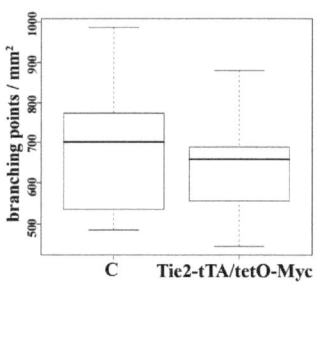
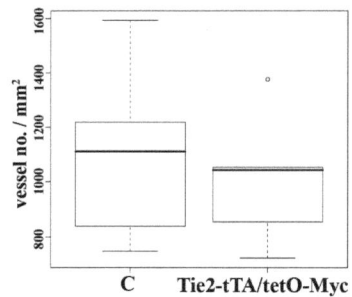
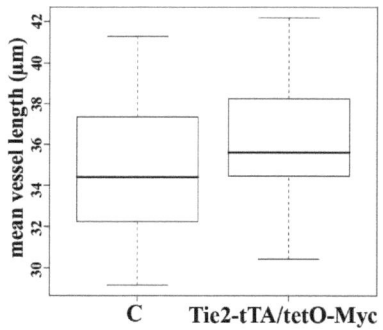

Figure 33. Morphometric analyses of back skin vascular architecture from control and Tie2-tTA/tetO-Myc embryos at E13.5.

Branching points per area, vessel number per area, and mean vessel length per image were quantified in control (wt or single transgenic) and Tie2-tTA/tetO-Myc embryos. N=6 animals and two images per animal were analyzed. The differences between the control and Tie2-tTA/tetO-Myc embryos in the analyzed parameters are not significant (Wilcoxon test).

3.3.6 Electron microscopical analysis of embryonic endothelium

To reveal further insights into the formation of the edema phenotype, electron microscopical characterization of the vasculature of Tie2-tTA/tetO-Myc embryos was performed.

Ultrastructural analyses of embryonic endothelium at E14.5 were done by transmission electron microscopy at the Institute of Pathology of the University of Tübingen by Prof. Hartwig Wolburg. Several embryonic organs, such as heart, lungs, liver, extremities, back skin, and brain were analyzed. On the basis of these results, the further studies were focused on the skin (on the extremities and back skin) of the embryos. It was found that wild type and single transgenic animals already showed a remarkable heterogeneity in blood vessel phenotype. The majority of vessels were intact (Figure 34a), and only few were obliterated and/or the surface of endothelial cells were enlarged (Figure 34b). A few endothelial cells were necrotic, or the circumference of the endothelial lining was discontinuous leading to local diapedesis of erythroblasts (data not shown). About 300 vessel profiles were inspected on multiple sections without any indication of apoptosis. In contrast, the blood vessels of Tie2-tTA/tetO-Myc double transgenic mice frequently revealed individual apoptotic cells interspersed within apparently normal endothelium (Figure 34c-f). 80% of vessel profiles showed single apoptotic figures. In other cases, 40% of the vessel profiles showed apoptotic figures, and the endothelial lining was completely destroyed resulting in a loss of contacts between individual cells (Figure 34d,e). In two out of seven Tie2-tTA/tetO-Myc embryos, an extreme condition was reached when endothelial cells were completely dissolved in apoptotic bodies (Figure 33f); such defects were never found in wild type or single-transgenic embryos.

Apoptosis of individual endothelial cells could well result in a breach of the vascular barrier and thus in a strongly increased permeability and an accumulation of fluid in the interstitium. It is noteworthy that endothelial tight junctions, which constitute the structural and functional elements of the vascular permeability barrier, remained unchanged in all parts of the vasculature of Tie2-tTA/tetO-Myc double transgenic embryos.

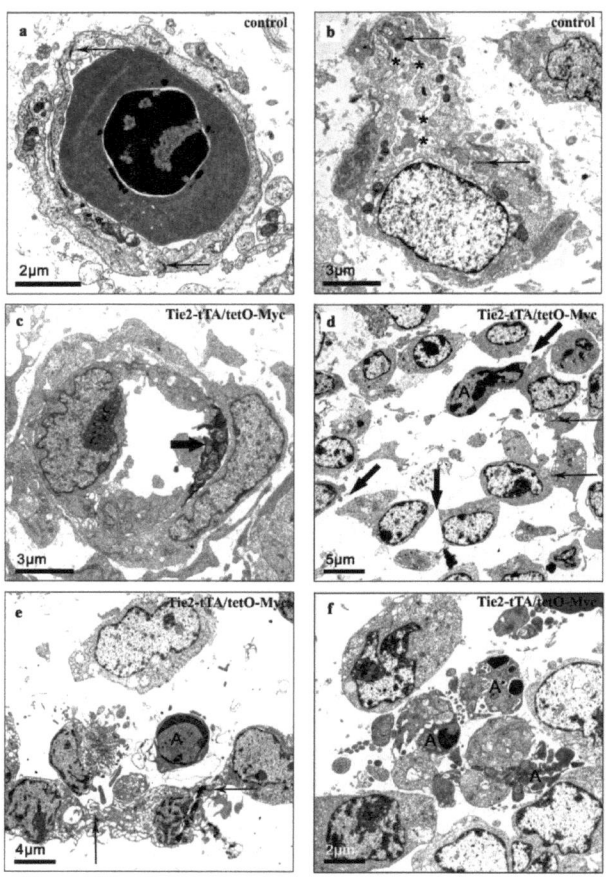

Figure 34. Electron microscopical analysis of skin blood vessels.

A, B: single transgenic, C-F double transgenic. A: a normal vessel with an erythroblast in its lumen. The endothelial cell sheet is continuous and interconnected by tight junctions (arrows). B: a blood vessel with apparently obliterated lumen (asterisks label the rest of the lumen), and increased surface. Endothelial cells are neither necrotic nor apoptotic and interconnected by tight junctions as well (arrows). C: Blood vessel with normal shape and continuous lining, but one single isolated apoptotic endothelial cell is shown (thick arrow). D: Blood vessel with a nearly completely dissociated lining. However, some tight junctions between endothelial cells are still found (thin arrows). Most endothelial cells are dissociated from each other leaving gaps between them (thick arrows). Some, but not all, endothelial cells are apoptotic (A). E: Higher degree of endothelial apoptosis (A) in a blood vessel revealing the typical segregation and condensation of nuclear chromatin. Some endothelial tight junctions are maintained confirming the presence of tight junction proteins. F: Culmination of endothelial apoptosis (A) evident by the blebbing of both nuclear and cytoplasmic apoptotic bodies. (Pictures taken and analysed by Prof. Hartwig Wolburg)

3.3.7 Quantification of endothelial cell apoptosis and proliferation

The ultrastructural analysis of the endothelium gave a hint that the endothelial cell-specific expression of c-Myc could induce apoptosis. This is not surprising given that several earlier studies had documented a critical role of Myc in regulating both apoptosis and cell proliferation (Askew 1991; Evan 1992; Evan 1998; Pelengaris 1999; Pelengaris 2002a; Pelengaris 2002b).

To closely assess these processes, I performed flow cytometry analyses of embryonic endothelial cells with specific markers for proliferation and apoptosis (Figure 35 and Figure 36). Proliferation of embryonic endothelial cells was measured by double staining with the CD31/PECAM-1 endothelial cell marker and with an antibody against Ki67 proliferating antigen. Flow cytometry analysis for endothelial cell apoptosis was done with CD31/PECAM-1 and cleaved caspase-3 double staining. Both measurements were quantified.

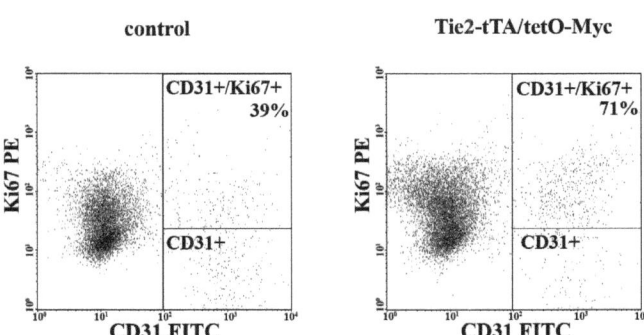

Figure 35. Analysis of embryonic endothelial cell proliferation.
FACS analysis of proliferating embryonic endothelial cells with double immunostaining for CD31/PECAM-1 and Ki67. Genotypes are indicated. A representative experiment is shown.

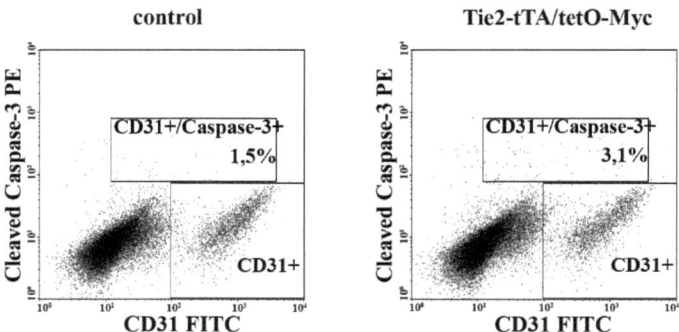

Figure 36. Analysis of embryonic endothelial cell apoptosis.
FACS analysis of apoptotic endothelial cells with double immunostaining for CD31/PECAM-1 and Caspase-3 for the indicated genotypes. A representative experiment is shown.

These analyses revealed a 1.5-2 fold increase of apoptotic endothelial cells in double transgenic embryos at E14.5 and E15.5, respectively (Figure 37B). Interestingly, there is a compensatory increase in the number of proliferating (Ki67-positive) cells (Figure 37A) at the same time points. The net result of increased apoptosis and proliferation is a largely unaltered absolute number of endothelial cells (Figure 37C), apparently at the expense of blood vessel integrity.

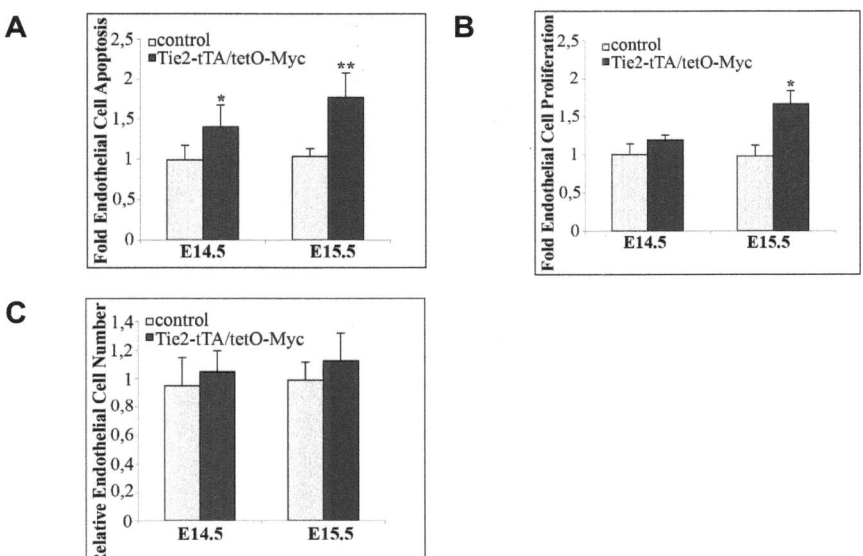

Figure 37. Quantification of embryonic endothelial cell apoptosis and proliferation.
Flow cytometry analysis was performed on cells from E14.5 and E15.5 embryos. (A) Quantification of endothelial cell apoptosis using double staining for cleaved caspase-3 and CD31/PECAM-1. In total, 26 control animals (wt or single transgenic, gray bars) and 11 double transgenic Tie2-tTA/tetO-Myc animals (black bars) were analyzed. Elevated apoptosis is found in Tie2-tTA/tetO-Myc embryos. *P<0.05, **P<0.005. (B) Quantification of endothelial cell proliferation using staining for the Ki67 antigen as proliferation marker and CD31/PECAM-1 staining to identify endothelial cells. Proliferation was measured in a total of 21 control mice (wt or single transgenic, gray bars) and 15 Tie2-tTA/tetO-Myc double transgenic animals (black bars). The proliferation rate is higher in Tie2-tTA/tetO-Myc embryos. *P<0.05 (C) Overall endothelial cell number does not change significantly. In total, 47 controls (gray bars) and 26 Tie2-tTA/tetO-Myc animals (black bars) were analyzed.

3.3.8 Isolation of embryonic endothelial cells

Our next aim was to detect the underlying molecular mechanisms of c-Myc overexpression in endothelial cells. Candidat target gene expression was analysed by quantitative RT-PCR. Therefore, embryonic endothelial cells were isolated.

Endothelial cells were purified by Fluorescence Activated Cell Sorting (FACS) as CD31/PECAM-1 and CD105/Endoglin double positive cells from Tie2-tTA/tetO-Myc double transgenic and control embryos at E14.5 (Figure 38). Flow cytometry analyses revealed that the purified $CD31^+/CD105^+$ cells were 92-98% positive for the CD31 endothelial surface marker (Figure 39).

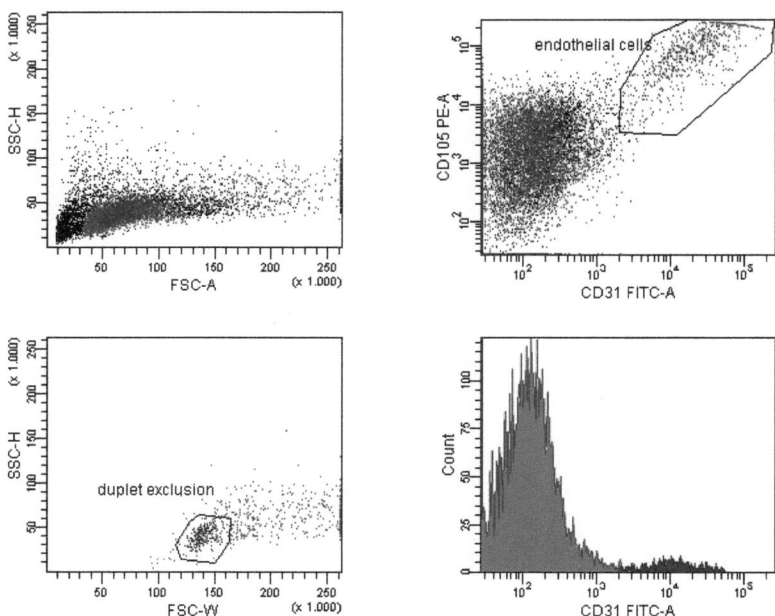

Figure 38. FACS embryonic endothelial cells.

The upper left panel with SSC-FSC gateshows all living cells. On the upper right panel endothelial cells were defined as CD31/PECAM-1 and CD105/Endoglin double positive population. After duplet exclusion with the definition of SSC-H/FSC-W gate (lower left panel), single endothelial cells of E14.5 embryos were sorted. The lower right panel shows the sorted endothelial cell population (blue).

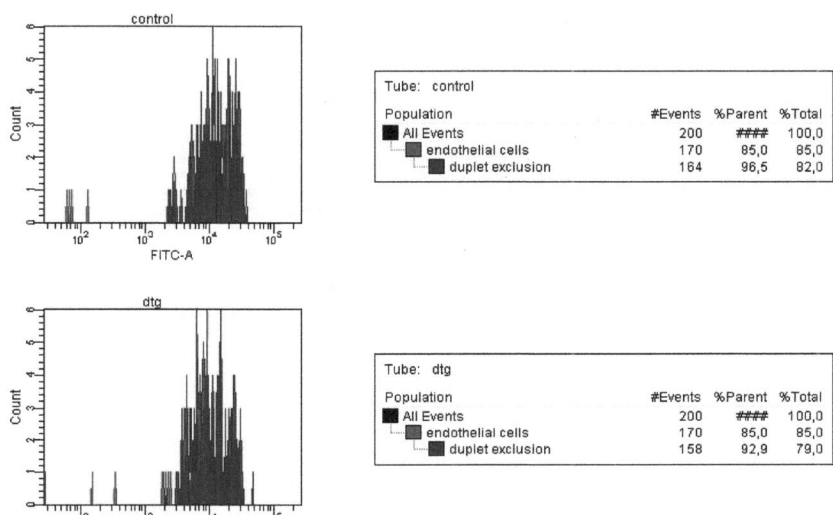

Figure 39. Purity of sorted embryonic endothelial cells.
The purity of E14.5 embryonic endothelial cells were determined by FACS analysis, by means of CD31/PECAM-1 expression. 92-98% of the sorted endothelial cells were positive for the CD31 endothelial surface marker. The upper panels show a representative control animal (wt or single transgenic). The lower panels show a representative double transgenic animal.

C-Myc-transgene expression was analyzed in the enriched endothelial cells by RT-PCR (Figure 40A) and detected specifically in the double transgenic endothelial cells only. Luciferase expression was assessed in the $CD31^-/CD105^-$ double negative and $CD31^+/CD105^+$ double positive cell fraction of control and Tie2-tTA/tetO-Myc embryos. Only low levels of luciferase-expression were detected in the $CD31^-/CD105^-$ double negative cell fractions of Tie2-tTA/tetO-Myc embryos (Figure 40B). As expected, Tie2-tTA/tetO-Myc embryos show high luciferase activity in the double positive cell fraction. Control animals have only background luciferase expression.

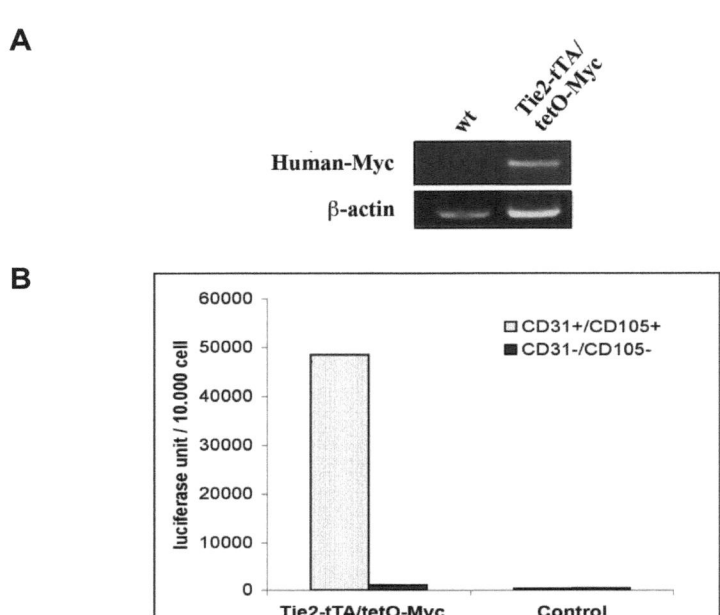

Figure 40. Transgene expression in purified endothelial cells.
(A) Expression of transgenic human-c-Myc was determined by RT-PCR in sorted endothelial cells from double transgenic and wild type control embryos. The endothelial cell preparation was 92-98% positive for the CD31 marker. C-Myc is expressed exclusively in double transgenic embryos. β-Actin was used as a reference gene. (B) Reporter gene expression is specific for the enriched endothelial cell population of double transgenic animals. Luciferase measurements were performed on 10'000 sorted cells. The double positive (CD31$^+$/CD105$^+$) endothelial cell fraction and the double negative (CD31$^-$/CD105$^-$) cell fraction from Tie2-tTA/tetO-Myc double transgenic as well as control embryos were analysed.

3.3.9 Angiogenic modulators expression in purified embryonic endothelial cells

Earlier experiments in c-Myc-deficient mice had revealed that c-Myc affects the expression of several angiogenic modulator genes. VEGF-A expression requires c-Myc activity, most likely via an indirect pathway (Baudino 2002). Therefore, quantitative RT-PCR analyses of candidate target genes in purified E14.5 embryonic endothelial cells were performed.

In endothelial cells derived from Tie2-tTA/tetO-Myc double transgenic embryos VEGF-A expression was upregulated more than two-fold (Figure 41A). Given that the levels of VEGF-A are regulated very tightly and that mutation of a single allele of VEGF-A results in early embryonic lethality the increased expression level of VEGF-A is of relevance (Carmeliet 1996; Ferrara 1996).

VEGF-A was originally discovered because its capacity to increase the permeability of microvessels to plasma and plasma proteins. Due to this finding it was named Vascular Permeability Factor (Senger 1983; Taylor AE 1984; Bates 1996). Therefore a two-fold overexpression of VEGF-A could well explain the edema phenotype of Tie2-tTA/tetO-Myc embryos.

Importantly, an increase in the levels of angiopoietin-2, a pro-angiogenic factor and potential antagonist of angiopoietin-1 was also observed. Ang-1 levels were only marginally affected leading to a shift in the net Ang-1/Ang-2 balance towards a prevalence of Ang-2. The expression of several other candidate genes that could explain the observed phenotype in vascular architecture and permeability, including the notch-ligand delta-like 4 (Dll4), was not altered.

Recent publications report the important role of adrenomedullin (AM) signaling during vascular development (Fritz-Six 2008; Ichikawa-Shindo 2008). Embryos lacking either AM, or the calcitonin-receptor like receptor (CRLR), the mediator of adrenomedullin signaling, or receptor activity modifying protein-2 (RAMP-2) develop severe edema and die at midgestation (Fritz-Six 2008; Ichikawa-Shindo 2008) Therefore, the expression of AM pathway signaling molecules in sorted endothelial cells was assessed and found to be down-regulated for RAMP-2 and upregulated for AM.

Given the increased numbers of sprouts, the expression of PDGF-B was analyzed, a marker of sprouting endothelial cells. No differences in PDGF-B expression between wild type and double transgenic embryos could be detected.

To control if the increased VEGF-A mRNA expression level was also reflected by an increased protein concentration, VEGF-A protein level was measured by ELISA. Wild type and Tie2-tTA/tetO-Myc embryonic lung tissue protein extracts were used for this experiment. This analysis revealed increased VEGF-A-levels in double transgenic embryos compared to wild types at all time points of embryonic development analyzed (Figure 41B).

A

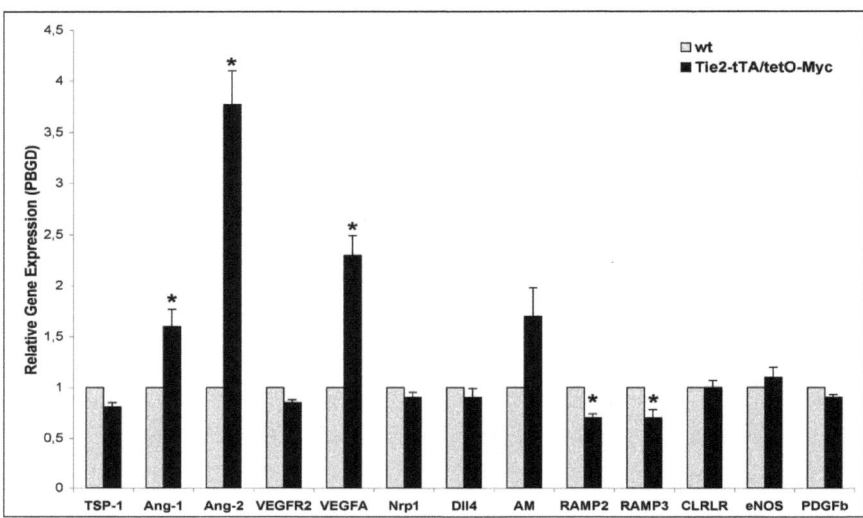

B

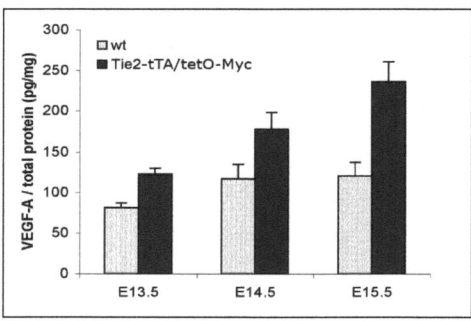

Figure 41. Expression of angiogenic modulators in sorted E14.5 endothelial cells and protein extracts

(A) Real-time quantitative PCR analyses of sorted endothelial cells from wild type (gray bars) and double transgenic Tie2-tTA/tetO-Myc embryos (black bars). VEGF-A, Ang-1, Ang-2 expression is significant upregulated and RAMP-2 and RAMP-3 expression is significant downregulated in double transgenic endothelial cells. SEM is shown as error bars. * $P < 0.05$ (B) VEGF-A expression was determined by ELISA measurements. The VEGF-A concentration was normalized against the total protein in the tissue sample. VEGF-A levels were increased in lungs from double transgenic Tie2-tTA/tetO-Myc embryos from day E13.5 onwards. SEM is shown as error bars.

3.3.10 Subsummary II

Endothelial c-Myc overexpression resulted in severe defects in the embryonic vascular system. Transgenic embryos suffer from widespread edema formation and multiple hemorrhagic lesions. They die between embryonic day E14.5 and E17.5.

The changes in vascular permeability are not caused by impaired blood vessel development. Blood vessels are covered with vascular basement membrane, pericytes, and smooth muscle cells, respectively. Tight junctions are well formed between endothelial cells. Analysis of the lymphatic vasculature of double transgenic embryos showed unaltered, normal developed lymphatic vessels. However, whole mount immunohistochemical analysis revealed alterations in the architecture of blood capillary networks. The dermal vasculature of c-Myc expressing embryos is characterized by a reduction in vessel branching and an increase in vessel sprouts at E14.5. Electron microscopic analysis revealed apoptotic endothelial cells. Moreover, the overall turnover of endothelial cells as measured by levels of apoptosis and proliferation is affected. Gene expression study of angiogenic modulators in c-Myc expressing embryos showed an elevated VEGF-A and Ang-2 expression.

Elevated expression of VEGF-A could be a cause of vascular hyperpermeability. To the edema formation could contribute the endothelial apoptosis due to the leaky vessels. The overexpression of c-Myc in endothelial cells induced a perturbed ratio of proliferation and apoptosis. Moreover, c-Myc upregulates the pro-angiogenic factors VEGF-A and Ang-2, which impairs vessel remodeling and results in aberrant vascular architecture.

4. DISCUSSION

4.1. C-Myc expression in endothelial cells of adult mice

4.1.1 C-Myc expression in Tie2-manner in adult mice

With the tetracycline regulatory system, the expression of c-Myc was induced postnatal in a Tie2-manner. Surprisingly, two different types of tumors were obtained in adult Tie2-tTA/tetO-Myc double transgenic animals.

The great majority of tumors developing were epithelial tumors. Only 15% of animals developed endothelial type tumors, angiosarcomas. Altough it is known that Tie receptors are almost exclusively expressed by vascular endothelial cells (Dumont 1994).

It was shown that in adult vasculature Tie2 is not only upregulated in areas of active angiogenesis but Tie2 is expressed and activated in the endothelium of all normal tissues (Peters 2004; Martin 2008). Tie2 was found to be expressed by a subpopulation of hematopoietic stem cells and bone marrow osteoblasts too. One additional cell population is also known to express Tie2. This cells are belonging to a subpopulation of monocytes, called Tie2-expressing monocytes (TEMs) (Augustin 2009). They seem to primarily account in tumors for the angiogenic activity of recruited tumor-associated macrophages (De Palma 2003; De Palma 2005; Augustin 2009).

To clearly define, how is possible that epithelial type tumors develop upon c-Myc expression in a Tie2-manner, we need further characterisation of these tumor cells by flow cytometry and immunohistochemistry.

4.1.2 C-Myc inactivation *in vivo*

The activation of oncogenes plays an important role in tumorigenesis. Strategies for the treatment of cancer that inactivate oncogenes are currently being developed. However, cessation of the pharmacologic inactivation of an oncogene may result in tumor regrowth. The isolation of tumor cell lines gives the possibility to characterize their transplantability and the treatment of growing tumors *in vivo*. Therefore, tumor cell lines were established from the tumors of Tie2-tTA/tetO-Myc mice. Further we were primarily interested whether oncogene inactivation after transplantation can induce tumor regression *in vivo*.

We demonstrated by treatment of angiosarcoma-bearing mice with doxycycline that the angiosarcomas were dependent on continuous c-Myc expression. Upon down-regulation of c-Myc expression, tumor regression was seen after 4 weeks of doxycycline treatment.

However, doxycycline treatment of a tumor cell line with endothelial characteristics derived from an angiosarcoma, did not stop cell proliferation although the tet-off system was switched off. We found that this cell line lost the transgene expression. This result is supported by the observations of Urban Deutsch, who found that isolated endothelial cells working under the control of the Tie2 promoter do not express the transgene *in vitro* anymore (unpublished data). Therefore, we tried to regain transgene expression under *in vivo* conditions in a xenograft model. After tumor cell line transplantation doxycycline treatment did not stop the tumor growth.

This phenomenon is already known from other c-Myc overexpressing systems (Felsher 1999a; D'Cruz 2001; Jain 2002). In a mammary carcinoma model, where c-Myc initiates and maintains the tumor, a subset of these tumor cells apparently escapes c-Myc dependence by activating endogenous *ras* oncogenes (D'Cruz 2001). In another *in vitro* myc-transformed fibroblast model c-Myc causes genomic instability leading to critical mutations and escapes the reversible cancer phenotype (Felsher 1999b).

4.1.3 Mouse model for human diseases

The two tumor types, angiosarcoma and adenoma, that developed in the Tie2-tTA/tetO-Myc mice are similar to human diseases and they even share some features with the corresponding human disease.

The dark purple or red nodular appearance of the tumor and spindle-like cell forms of the murine angiosarcoma are similar to Kaposi's sarcoma characteristic in human. In humans Kaposi's sarcoma is highly vascular, containing abnormally dense and irregular blood vessels, which leak red blood cells into the surrounding tissue and give the tumor its dark color. Kaposi's sarcoma lesions contain tumor cells with a characteristic abnormal elongated shape, called spindle cells.

Kaposi's sarcoma lesions are typically found on the skin. Usually affected areas include lower limbs, face and genitalia, but spread elsewhere is also common. Tie2-tTA/tetO-Myc mice showed nodules on the extremities and mouth.

Kaposi's sarcoma in human is caused by a virus infection through human herpesvirus 8 (HHV8 or Kaposi's sarcoma-associated herpes virus, KSHV). It is a common disease associated with AIDS infection and immunosuppression. The endemic form of Kaposi's sarcoma is known in young African people. Classical Kaposi's sarcoma was described in elderly men from Mediterranean region, and of Eastern European descent.

The other type of tumor classified as adenoma, occurs in humans and rodents as well. Adenomas are benign epithelial tumors of glandular (endocrine and exocrine), mucosal (stomach, intestine) and ductae origin. Over time they may progress to become malignant, a point at which they are called adenocarcinomas.

4.1.4 Animal model for Kaposi's sarcoma

Many efforts have been made to develop mouse models of Kaposi's sarcoma. While some of these models used dimethyl hydrazine to induce colonic cancer and angiosarcoma, another model was based on the Tie2 vascular endothelial cell-specific promoter similar to our study (Dube 1992; Madarnas 1992; Montaner 2003).

Gutkind and colleagues engineered transgenic mice to express the avian retroviral receptor, TVA, under the control of the Tie2 promoter. Tie2-tva animals were infected with RCAS-vGPCR (G-protein-coupled receptor) avian retrovirus, encoding an open reading frame (ORF 74) of Kaposi's sarcoma-associated herpes virus. Signaling by this KSHV G-protein-coupled receptor results in cell transformation and tumorigenicity, and induces a switch to an angiogenic phenotype mediated by VEGF-A (Bais 1998). Infected Tie2-tva mice developed visible vascular tumors beside paw, tail and dermis, involving multiple internal organs too for example intestine, liver, heart, and peritoneum.

In contrast, Tie2-tTA/tetO-Myc mice developed vascular tumors on the extremities, tail and mouth but not on internal organs. An explanation for this phenomenon could be that the Tie2-tva mice were infected by intraperitoneal injection of RCAS- vGPCR virus. Thus they could develop the multiple internal organ vascular tumors.

4.2. C-Myc expression in embryonic endothelial cells

4.2.1 Possible origins of vascular permeability

Tie2-tTA/tetO-Myc embryos, which overexpress c-Myc in endothelial cells, showed embryonic lethality accompanied by severe edema and a hemorrhagic phenotype. These symptoms can arise due to a primary defect deriving from blood or lymphatic vessels or secondarily due to organ failure. We focused our research on primary vascular defects and extensively analysed them in the recent work.

Vascular leakage could be the outcome of a not completely or defective developed blood or lymphatic vessels. Different defects can be the reason for the affected barrier function. One reason is the lack of vascular mural cells: pericytes on microvessels, SMCs on arteries. This phenomenon could result in an unstable, fragile vessel structure and contribute to edema. PDGF-B is expressed on immature and growing endothelial cells, it recruits pericytes which express the PDGFRβ receptor (Armulik 2005). Both the PDGF-B and PDGFRβ -deficient mouse embryos were found to lack microvascular pericytes and they developed lethal hemorrhages and edema in late embryogenesis (Lindahl 1997; Hellstrom 1999; Hellstrom 2001). In our analyses of vessel morphology we could show that Tie2-tTA/tetO-Myc embryonic capillaries have an intact pericyte cover. Further quantitative RT-PCR analyzes of PDGF-B expression, in sorted endothelial cells did not show differences between Tie2-tTA/tetO-Myc and wild type embryos. We also analyzed the SMC cover of blood vessel walls. However, the pattern and number of SMCs of Tie2-tTA/tetO-Myc blood vessels was comparable to control embryos.

Another reason of defective barrier function can be the excision of extracellular matrix proteins. Extracellular matrix proteins contribute to the maintenance of the normal endothelial barrier due to their influence on the trans-endothelial flux of solutes and fluids (Lum 1994). *In vitro* studies show that remodeling of the extracellular matrix by proteases increased the permeability of the barrier (Partridge 1992). *In vivo* matrix metalloproteinase-9 (MMP-9 or gelatinase B) expression induced edema formation in rabbit lungs (Passi 1999). Interestingly, it was shown that c-Myc induces MMP-9 expression in endothelial cells (Magid 2003). Therefore we assessed MMP-9 expression in our mice. Only a 1,5 fold increase of MMP-9 protein expression level was noticed in lung extracts of double transgenic embryos. We suppose that this level of MMP-9 expression is representative for the endothelium in the whole embryonic body. Therefore this slight increase of MMP-9 expression most likely cannot explain the formation of the edema in the double transgenic embryos.

A third reason, which can result in leaky vessels, is developmental defects in basement membrane assembly. For example, the lack of a basement membrane component, heparan sulfate proteoglycan, perlecan in mice results in fetal lethality between 10 to 12 days of gestation due to bleeding in the pericardial sac (Arikawa-Hirasawa 1999; Costell 1999). Therefore it was reasonable to analyze the integrity of the basement membrane in our mice. We assessed the expression of laminin, a basement membrane component, in embryonic skin samples. This analysis showed equal assembly of the basement membrane of Tie2-tTA/tetO-Myc and control embryos.

Another possible source of edema could be the dysfunction or lack of inter-endothelial junctions. Tight junction proteins are needed for the maintenance of endothelial barrier properties and adherens junction components are necessary for correct vascular morphogenesis (Nyqvist 2008). We extensively analysed the expression of tight junction members.

Adherens junction components are necessary for correct vascular morphogenesis. Defect in expression of AJ proteins could result only as a secondary effect in endothelial barrier dysfunction. Therefore AJ components were not further analysed in my work.

The integrity of TJ is regulated by claudins, occludin and JAMs. Only claudin-5 out of the 24-member claudin family, is expressed in endothelial cells. Inactivation of the tight junction component *Claudin5* gene in mice did not morphologically alter the vascular network or the ultrastructural appearance of the TJs (Nitta 2003). However, claudin-5-deficient pups died within 10 h of birth due to a size-selective loosening of the blood–brain barrier against molecules that were smaller than 800 Da (Nyqvist 2008).

The transmembrane protein occludin is present in endothelial cells, and it is expressed at highest level in endothelial cells of the central nervous system (Nyqvist 2008). However, no effects on vascular morphology or blood brain permeability have been reported in mice that lack occludin (Saitou 2000). *In vitro* data showed that deletion of NH2-terminal region of occludin causes TJ-leakiness (Bamforth 1999). Therefore the NH2 terminus is crucial for regulating barrier function (Mehta 2006).

JAM-A (junctional adhesion molecule-A) gene inactivation in mice did not affect the development of the vascular system in the embryo (Cera 2004; Cooke 2006). Therefore I did not analyse further its expression in my work.

ZO-1 and ZO-2 (zonula occludens) are intracellular components of TJs. Ablation of *ZO-2* resulted in arrested development at E5.5 and lethality before the onset of vascular development (Xu 2008). By contrast, embryos deficient for *ZO-1* developed normally until E8.5, and thereafter displayed growth retardation, defective chorioallantoic fusion and yolk-sac angiogenesis (Katsuno 2008). Blood vessels in ZO-1-null yolk sacs were characterized by defective remodeling and compartmentalization, as they expanded with only a few adhesion sites.

We closer analysed the above-mentioned tight junctional proteins in dermal blood vessels of E14.5 embryos. However, did not find an impaired expression pattern of these proteins between Tie2-tTA/tetO-Myc and control embryos. The normally developed tight junctions were approved by electron microscopy images, where tight junctions remained unchanged in the vasculature of double transgenic embryos.

Therefore, the leaky vessel phenotype of double transgenic embryos can not be explained by tight junctional barrier defects.

Defects of the lymphatic vasculature can cause severe edema symptom. Defects involve the deficient development of lymphatic vessels and not sufficient drainage of lymph.
The gross structure of lymphatic vessels was analyzed in double transgenic and control mice but obvious alterations were not detected.
Interestingly, in our endothelial cell-specific gene expression study, we found an elevated expression of VEGF-A. In the literature it was reported that VEGF-A overexpression in mouse skin induced enlargement of lymphatic vessels (Nagy 2002; Hirakawa 2005). Several other studies suggest that VEGF-A/VEGFR-2 signaling may also directly affect the lymphatic endothelium (Makinen 2001b; Karpanen 2006). Therefore, we inspected the size of lymphatic vessels of double transgenic embryos. However, dermal lymphatic vessels of the double transgenic embryos were normal, not enlarged.

Vascular permeability induced by VEGF-A

VEGF-A is not only essential during development of the vasculature but was found as a permeability-increasing agonist specific for endothelial cells (Vandenbroucke 2008). Dvorak and colleagues originally discovered VEGF-A and called it Vascular Permeability Factor (Senger 1983). They observed that tumor cells secrete VEGF-A, which induces vascular leaks in the microvasculature. Later studies revealed a molecular mechanism how VEGF-A induces vascular permeability. VEGF-A phosphorylates the adherens junction proteins VE-cadherin and β-catenin and induces their dissociation, which breaks down the link between the interendothelial junctions and the actin cytoskeleton (Esser 1998; Weis 2008).
The levels of VEGF-A expression during embryogenesis need to be tightly regulated. It has been shown that two- to threefold overexpression of VEGF-A results in severe abnormalities in embryonic heart development, edema formation and embryonic lethality at E12.5-E14 (Gerber 1999; Miquerol 2000). On the other hand, inactivation of a single VEGF-A allele in mice already resulted in embryonic lethality between E11 and E12 (Carmeliet 1996; Ferrara 1996).
Previously, it has been reported that VEGF-A production is reduced in *c-myc -/-* ES cells as well as in *c-myc -/-* embryos. Baudino and colleagues proposed an indirect mechanism of *Vegfa* gene regulation by c-Myc (Baudino 2002).
In our study, expression levels of VEGF-A were found to be elevated significantly both at the mRNA and protein levels in Tie2-tTA/tetO-Myc endothelial cells. These results mirror the earlier report of Baudino as we see a comparable increase of VEGF-A levels in c-Myc overexpressing endothelial cells. Thus, we conclude that the elevated VEGF-A expression

induced by c-Myc in Tie2-tTA/tetO-Myc embryos could impair the vascular barrier function and contribute to the increased vascular permeability.

4.2.2 Vascular permeability protecting factors

The stabilization and destabilization of endothelial cell adhesion is known to be balanced by the counteraction of Ang-1 and Ang-2 on the Tie2 receptor (Augustin 2009; Thomas 2009).

It was reported that transgenic overexpression of Ang-2 disrupts blood vessel formation in the mouse embryo, which leads to midgestational lethality (Maisonpierre 1997). In contrast, it was shown that Ang-1 inhibits vascular leakage by strengthening inter-endothelial junctions (Gamble 2000; Fukuhara 2009). Moreover, *in vivo*, Ang-1 overexpression and/or systemic delivery decreased vessel leakiness in response to permeability-inducing agents like VEGF-A (Thurston 1999; Thurston 2000).

Tie2-tTA/tetO-Myc embryos had a 1.5 fold increase in the expression levels of Ang-1 and Ang-2 expression level was more than 3.5 fold increased. It is possible that the elevated expression of Ang-1 is a secondary effect. A permeability-decreasing event in order to balance the Ang-2 and VEGF-A induced vascular permeability.

Recent publications report the important role of adrenomedullin (AM) signaling during vascular development.

AM was found as a vasodilator peptide that is pathologically elevated in a variety of tumors and cardiovascular conditions. In addition, AM treatment induces proliferation, migration, and capillary tube formation of cultured human umbilical vein endothelial cells (HUVEC) (Miyashita 2003a; Miyashita 2003b; Fernandez-Sauze 2004). AM signals through calcitonin-receptor like receptor (CRLR) which ligand-bindig affinity can be changed by interaction with receptor activity modifying proteins (RAMPs). In the mammalian vasculature three different RAMPs, RAMP-1, -2, and -3, are highly expressed (Fritz-Six 2008). Embryos lacking either AM, CRLR, or RAMP-2 develop severe edema and die at midgestation (Caron 2001; Dackor 2006; Fritz-Six 2008; Ichikawa-Shindo 2008). In these mice the embryonic lethality arises due to combined defects in both blood and lymphatic vessels (Kahn 2008).

We tested the expression of AM pathway signaling molecules in sorted endothelial cells and found that RAMP-2 is slightly down-regulated and AM expression is enhanced in Tie2-tTA/tetO-Myc embryos. However, it was recently reported that AM stabilizes lymphatic endothelial cell-cell junctions and thereby prevents the VEGF-A-mediated increased permeability *in vitro* (Dunworth 2008). AM-treatment reorganizes the endothelial junctional proteins ZO-1 and VE-cadherin and therefore a tighter paracellular seal is formed between cells. With regard to blood endothelial permeability, AM treatment of HUVECs reduces the inflammatory mediators induced hyperpermeability (Hippenstiel 2002; Hocke 2006). Therefore, we can assume that the elevated expression of AM in Tie2-tTA/tetO-Myc embryos could result in stabilized endothelial cell-cell junctions. This idea is supported by the immunofluorescence analysis of tight junctional molecules in embryos, where we did not see any signs of leakiness between TJs of double transgenic embryos. Importantly, Tie2-tTA/tetO-Myc embryos show a normal developed lymphatic vasculature, suggesting that the reduced AM signaling did not affect lymphatic vessel development.

4.2.3 Factors modulating vessel remodeling

In addition to the edema, the Tie2-tTA/tetO-Myc embryos, exhibited a striking phenotype in their vascular architecture. Therefore we were interested in the question, which changes in the expression of distinct molecular modulators controlling vessel remodeling and vessel sprouting affected the impaired blood vessel assembly in double transgenic embryos.

Angiopoietin-2 and VEGF-A

The important role of angiopoietin-2 in the regulation of vascular remodeling depends on the local cytokine milieu (Augustin 2009; Thomas 2009). In the presence of VEGF-A, Ang-2 cooperates with VEGF-A to promote sprouting, proliferation, and migration of endothelial cells. By contrast, in the absence or inhibition of VEGF-A, Ang-2 up-regulation promotes vascular destabilization and subsequent vessel regression (Lobov 2002; Fiedler 2006). Transgenic overexpression experiments of Ang-2 alone or Ang-2 in combination with VEGF-A showed the cooperation between this two angiogenic modulators. This experiments revealed that a high Ang-2 to VEGF-A ratio leads to vessel regression, whereas high VEGF-A levels in the presence of Ang-2 results in angiogenesis (Oshima 2004; Oshima 2005).

Tie2-tTA/tetO-Myc embryos have 2 fold elevated amounts of VEGF-A and an even further elevated, 3,5 fold expression of Ang-2. The double transgenic embryos did not show an increased angiogenesis. The elevated expression of VEGF-A and Ang-2 is accompanied by an inadequate vascular remodeling, which results in the establishment of an aberrant blood vessel architecture.

Notch-signaling and VEGF-A

During angiogenic expansion, VEGF-A stimulates endothelial cells to sprout and proliferate to form new vessel structures. Several recently published studies have demonstrated that Notch-signaling regulates the sprouting and branching behavior in vessels by influencing the differentiation, migration and proliferation of vascular tip cells (Noguera-Troise 2006; Ridgway 2006; Hellstrom 2007; Lobov 2007; Scehnet 2007; Suchting 2007).

Tip cells are distinct and functionally specialized microvascular endothelial cells (Gerhardt 2003). Their place is at the most terminal region in a vascular sprout. They are followed by stalk cells, which form the stalk during the sprouting process (Gerhardt 2008). The Notch-ligand Delta-like-4 (Dll-4) is expressed in tip cells (Claxton 2004) and its receptor Notch-1 is expressed in stalk cells (Hofmann 2007). VEGF-A induces Dll4, which functions to pattern the endothelial population into tip and stalk cells. The tip cells then migrate along the VEGF-A gradients, whereas stalk cells proliferate in a polarized fashion to supply further endothelial cells. The combined effect of migration by tip cells and proliferation by stalk cells results in vascular growth (Gerhardt 2003). VEGF-A gradients, arising through regulated retention of VEGF-A on the cell surface and in the extracellular matrix, govern these processes (Gerhardt 2008). However it was found that reduced Notch-signaling results in increased numbers of tip cells, filopodia extension, and vessel branching (Gridley 2007).

In our vessel architecture study, we observed elevated numbers of vascular sprouts and fewer vessel branches in the dermis of double transgenic embryos. To investigate this phenomenon at the molecular level, Dll-4 and PDGF-B sprout marker expression was analysed in sorted embryonic endothelial cells. Their expression levels were comparable in Tie2-tTA/tetO-Myc and control embryos. Consequently, c-Myc overexpression did not reduce Notch-signaling.

However, in this experiment the expression of the sprout markers was tested in a pool of sorted endothelial cells. It would be interesting to isolate only tip cells and look for the proper expression of these sprout markers.

Therefore, sprouts could be a result of an elevated angiogenic activity triggered due to elevated VEGF-A expression. Alternatively, sprout-like figures could arise as result of vessel involution caused by endothelial cell apoptosis.

4.2.4 C-Myc and angiogenic switch

A direct consequence of certain oncogenic lesions in the vasculature is angiogenesis (Evan 2001). C-Myc's angiogenic capacity has been observed in skin, lymphoma, neuroblastoma and in a fibroblast xenograft model (Pelengaris 1999; Brandvold 2000; Breit 2000; Ngo 2000). Inactivation of c-Myc in inducible Myc-knock out models leads to rapid regression of tumor vasculature triggering concomitant tumor involution (Fotsis 1999; Okajima 2000; Pelengaris 2002b). C-Myc directly induces angiogenesis in part by induction of VEGF-A and downregulation of the angiogenesis negative modulator thrombospondin-1 (TSP-1) (Janz 2000; Ngo 2000). This process is called angiogenic switch (Volpert 2003). In most of the cases the newly developed tumor vasculature is leaky, immature and unstable.

We were interested, whether c-Myc induces TSP-1 downregulation also in the developing embryo. However, mRNA expression level analysis of TSP-1 by Tie2-tTA/tetO-Myc embryos did not reveal a significant decrease. Therefore, we could not find the evidence of angiogenic switch by Tie2-tTA/tetO-Myc embryos.

4.2.5 C-Myc: apoptosis and/or proliferation?

In our study we have demonstrated that endothelial cell-specific expression of c-Myc influences both the cell cycle progression and apoptosis. The transmission electron microscopy study revealed an accumulation of interspersed apoptotic endothelial cells in blood vessel walls. FACS analyses of endothelial cells proved a significant elevation in apoptosis at E14.5 and E15.5, and showed an increased proliferation rate starting at E14.5.

It was already described that the mitogenic and proapoptotic properties of c-Myc are genetically inseparable. Although the processes of cell renewal and cell death appear to be opposing and mutually contradictory, substantial evidence indicates that the two are linked (Evan 1998). Ectopic expression of c-Myc is sufficient to drive many cells into the cell cycle even in the absence of external mitogens. However, c-Myc also promotes apoptosis, although the precise mechanisms by which this occurs have not been completely elucidated (Askew 1991; Evan 1992; Evan 1998).

One possible way to induce apoptosis is due to the Arf tumor suppressor which is a target gene of c-Myc (Zindy 1998). Induction of Arf, an alternate product of the INK4a locus, triggers upregulation of p53 through its inhibitory action on MDM-2 (Sherr 2000). The p53 tumor suppressor functions as a sequence-specific transcription factor responsive to a wide array of signals that stress the cell, including DNA damage, hypoxia or hyperproliferative signals induced oncogenes such as c-Myc (Nilsson 2003). The Arf-p53 pathway is an important mediator of c-Myc-induced apoptosis, as loss of either *p53* or *Arf* impairs c-Myc-induced apoptosis (Hermeking 1994; Wagner 1994; Zindy 1998). To prove that c-Myc induces the endothelial cell apoptosis via the Arf-MDM2-p53 pathway, we examined the expression level of Arf in our endothelial-cell-specific mRNA-expression study. However, an elevated mRNA expression of Arf could not be detected in Tie2-tTA/tetO-Myc embryonic endothelial cells.

Caspases are the effector molecules in apoptosis. C-Myc induces the release of cytochrome c from the mitochondria during apoptosis. Once released into the cytoplasm, cytochrome c associates with Apaf1 (Apoptotic protease-activating factor 1) to create the apoptosome, a complex that activates procaspase-9. Caspase-9 is autocatalytically activated and activates a downstream caspase-effector cascade that predominantly involves caspase-3. This results in proteolytic cleavage of cellular components leading to the demise of the cell (Pelengaris 2002a). To evaluate the c-Myc induced apoptosis through caspases, the expression of activated caspase-3 was quantiatively analysed in embryonic endothelial cells. An elevated cleaved caspase-3 level was measured by E14.5 and E15.5 Tie2-tTA/tetO-Myc mice.

The neoplastic impact of any oncogenic mutation is likely to lead to dramatically different outcomes depending on the context. Cell-type-specific levels of endogenous pro- and anti-apoptotic effectors, together with microenvironmental factors and other oncogenic events, all influence the signalling flux through pathways that contribute to cell proliferation and viability (Lowe 2004). For example, acute activation of c-Myc in pancreatic beta-cells leads to rapid beta-cell involution and diabetes (Pelengaris 2002b). By contrast, activation of c-Myc in skin triggers proliferation without cell death, probably because of an abundance of local survival factors, resulting in rapid development of papillomatous hyperplasias (Pelengaris 1999). Therefore, it is not unexpected that endothelial cell-specific expression of c-Myc results in elevated cell cycle progression and at the same time increased levels of apoptosis. Although these two effects are balanced with no overall loss of endothelial cells, these processes could lead to transient impairment of endothelial integrity that might contribute to the observed vascular leakiness.

4.2.6 The role of c-Myc in embryonic vascular development

Earlier *in vivo* studies on mice with systemic *c-myc* deletion showed the crucial role of c-Myc during embryonal development (Davis 1993). C-Myc knock-out mice were lethal between E9.5 and E10.5 of gestation. The embryos were smaller in size and retarded in development compared with their littermates. The pathologic abnormalities include circulatory and neural tube developmental defects. The multi-organ failure was believed to underlie the early developmental arrest (He 2008).

Later a closer study on the c-Myc -/- embryos showed further abnormalities in vascular and hematopoietic development (Baudino 2002). It was found that loss of c-Myc results in impaired vasculogenesis, angiogenesis, primitive erythropoiesis and impaired angiogenic modulator expression.

Recent reports focus on the role of c-Myc during vascular and hematopoietic development in different cell types (Dubois 2008; He 2008). Two different knock-out mouse models were used to target cell-specific c-Myc deletion.

Tie1-Cre;c-myc$^{flox/-}$ mice have a c-Myc knock-out in the majority of endothelial cells and in a subset of hematopoietic cells. Analysis of these mutants revealed that deletion of c-Myc results in the development of 80% of endothelial cells which still allows the formation of a largely normal vascular system.

Tie2-Cre;c-myc$^{flox/-}$ mutants have a c-Myc deletion in hemangioblasts, the common mesoderm-derived precursors of hematopoietic and endothelial lineages. When c-Myc is deleted in these common progenitors, angiogenesis and hematopoiesis are severely affected. Tie2-Cre;c-myc$^{flox/-}$ mutants are embryonically lethal around E11.5. They show normal vasculogenesis, but a defective capillary remodeling and hematopoiesis. Tie2-Cre;c-myc$^{flox/-}$ mutants reveal that c-Myc is not required for vasculogenesis, but that it is crucial for hematopoiesis and has an important role in the control of angiogenesis.

In our study, we used the Tie2 promoter for the conditional expression of c-Myc in developing endothelial cells. C-Myc-overexpressing embryos developed a normal primary blood vessel structure. However, overexpression of c-Myc in endothelial cells resulted in severe defects in angiogenesis and embryonic lethality between E14.5-E17.5. Vascular defects were associated with increased vascular permeability and alterations in vascular architecture.

In both mouse models, the overexpression or deletion of c-Myc in Tie2-manner resulted in a normal vasculogenesis. However, in both cases angiogenesis was impaired and capillary remodeling was defective.

It is important to point out that the lethal phenotype by day E11.5 of Tie2-Cre;c-myc$^{flox/-}$ mutants is a combination of defects in both hematopoietic and endothelial cells. In later stages of embryonic development, Tie2 expression is restricted to endothelial cells. Hematopoietic cells lose Tie2 expression. Thus, we suppose that in our model the embryonic midgestational lethality is exclusively due to the influence of c-Myc in developing endothelial cells.

4.2.7 Proposed model of the role of c-Myc during embryonic vascular development

In Tie2-tTA/tetO-Myc embryos, the vasculature established during vasculogenesis and angiogenesis, appeared normal until E13.5. Blood vessels in edematous areas of the embryonic skin at E14.5 still had normal vessel morphology with fully developed basement membranes, extracellular matrix, and an intact pericyte and smooth muscle cell cover. However, during consecutive blood vessel formation, c-Myc affects expression of angiogenic regulators and endothelial cell-turnover (Figure 42A,B). Transmission electron microscopy revealed accumulations of interspersed apoptotic endothelial cells, which could at least in part explain the interruption of vascular integrity and increased vessel permeability. Importantly, tight junctions were intact between endothelial cells, indicating that tight junctional failures did not critically contribute to vascular leakage. Elevated expression of VEGF-A as vascular permeability factor could be a cause of the vascular hyperpermeability and edema formation. Moreover, the elevated expression of VEGF-A together with Ang-2 in endothelial cells is accompanied by inadequate vascular remodeling, which results in the establishment of an aberrant blood vessel architecture with ongoing angiogenic activities.

We propose that the leaky vascular phenotype is the consequence of a perturbed ratio of proliferation and apoptosis induced by c-Myc overexpression in endothelial cells. C-Myc-induced expression of the pro-angiogenic factors VEGF-A and Ang-2 impairs vessel remodeling and results in aberrant vascular architecture and immature vasculature with ongoing angiogenic activities.

A

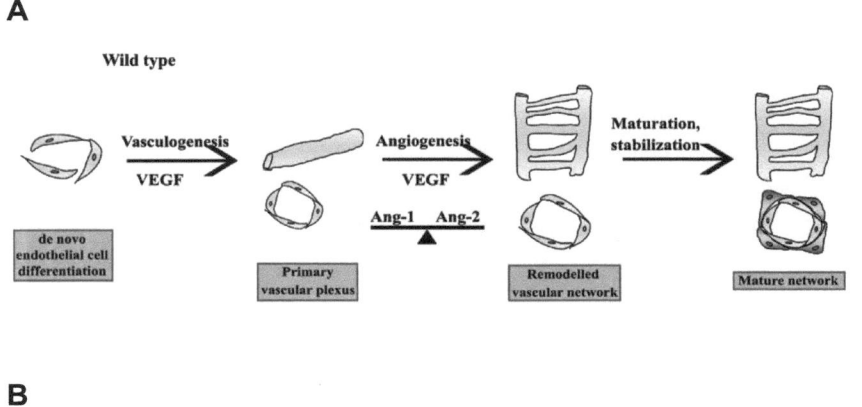

B

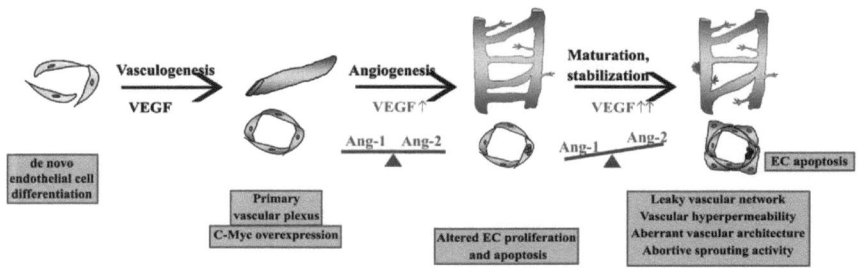

Figure 42. Model illustrating the role of c-Myc during embryonic vascular development.
(A) In wild type embryos, vasculogenesis, angiogenesis, vessel maturation, and stabilization occurs in a highly coordinated manner. These processes are controlled by the balanced expression of angiogenic modulators. (B) In the presence of excess c-Myc activity in endothelial cells vasculogenesis and angiogenesis occur normally up to E13.5. However, during consecutive blood vessel formation c-Myc affects expression of angiogenic regulators and endothelial cell-turnover. Elevated VEGF-A and Ang-2 levels contribute to vascular hyperpermeability and edema. In addition, endothelial cell apoptosis results in a leaky vascular network and finally hemorrhages. Moreover, elevated expression of pro-angiogenic factors VEGF-A and Ang-2 impairs vessel remodeling and results in aberrant vascular architecture and immature vasculature with ongoing angiogenic activities.

ABBREVIATIONS

AJ	adherens junction
AM	Adrenomedullin
Ang	Angiopoietin
AS	angiosarcoma
Apaf1	Apoptotic protease-activating factor 1
bFGF	basic fibroblast growth factor
BG	Bulbourethral gland
BR/HLH/LZ	basic-region/helix-loop-helix/leucine-zipper
CRLR	calcitonin-receptor like receptor
CTD	carboxy-terminal domain
Dll4	Delta-like 4
DOX	Doxycycline
DSBs	DNA double-strand breaks
Dtg	double transgenic
E	Embryonic developmental age
ECM	extracellular matrix
ELISA	Enzym Linked Immunoadsorbent Assay
FACS	Fluorescence Activated Cell Sorting
FITC	Fluoresceinisothiocyanat
HIF-1	Hypoxia-inducible factor-1
HSPGs	Heparan sulphate proteoglycans
HUVEC	Human Umbilical Vein Endothelial Cells
KSHV	Kaposi's sarcoma-associated herpes virus
ILM	Institute for Laser Technology
INR	initiator element
IVC	individually ventilated cages
JAMs	junctional adhesion molecules
LEC	lymphatic endothelial cell
LYVE-1	Lymphatic endothelial hyaluronan receptor-1
MDM2	Murine double minute 2
MMPs	Membrane Metalloproteases
mRNA	messenger RNA

MBI-III	Myc Boxes
Np	Neuropilin
NTD	amino-terminal domain
ODC	ornihine decarboxylase
PC	pericytes
PDGF	Platelet-derived growth factor
PE	Phycoerythrin
RAMP-2	Receptor activity modifying protein-2
SHMT	serine-hydroxymethyl-transferase
SMC, vSMC	vascular smooth muscle cell
SLC	secondary lymphoid chemokine
T	thymus
TAD	transactivation domain
TEMs	Tie2-expressing monocytes
tetO	tetracycline responsive minimal promoter
TGFβ-1	Transforming growth factor-β1
TJ	tight junction
TSP-1	Thrombospondin-1
tTA	tetracycline-transactivator
VE-cadherin	Vascular Endothelial-cadherin
VEGF	Vascular Endothelial Growth Factor
VEGFR	Vascular Endothelial Growth Factor Receptor
VPF	Vascular Permeability Factor
WPB	Weibel-Palade body
wt	wild type
ZO	zonula occludens

REFERENCES

Aalami, O. O., D. B. Allen, et al. (2000). "Chylous ascites: a collective review." Surgery **128**(5): 761-78.

Adhikary, S. and M. Eilers (2005). "Transcriptional regulation and transformation by Myc proteins." Nat Rev Mol Cell Biol **6**(8): 635-45.

Alberts, J. A., Lewis J, Raff M, Roberts K, Walter P (2001). "Molecular Biology of the Cell." **4th**.

Alitalo, K. and P. Carmeliet (2002). "Molecular mechanisms of lymphangiogenesis in health and disease." Cancer Cell **1**(3): 219-27.

Alon, T., I. Hemo, et al. (1995). "Vascular endothelial growth factor acts as a survival factor for newly formed retinal vessels and has implications for retinopathy of prematurity." Nat Med **1**(10): 1024-8.

Arikawa-Hirasawa, E., H. Watanabe, et al. (1999). "Perlecan is essential for cartilage and cephalic development." Nat Genet **23**(3): 354-8.

Armulik, A., A. Abramsson, et al. (2005). "Endothelial/pericyte interactions." Circ Res **97**(6): 512-23.

Askew, D. S., R. A. Ashmun, et al. (1991). "Constitutive c-myc expression in an IL-3-dependent myeloid cell line suppresses cell cycle arrest and accelerates apoptosis." Oncogene **6**(10): 1915-22.

Augustin, H. G., G. Y. Koh, et al. (2009). "Control of vascular morphogenesis and homeostasis through the angiopoietin-Tie system." Nat Rev Mol Cell Biol **10**(3): 165-77.

Baddeley, A., Vedel Jensen EB (2004). Stereology for Statisticians. London, Chapman & Hall.

Bais, C., B. Santomasso, et al. (1998). "G-protein-coupled receptor of Kaposi's sarcoma-associated herpesvirus is a viral oncogene and angiogenesis activator." Nature **391**(6662): 86-9.

Baluk, P., J. Fuxe, et al. (2007). "Functionally specialized junctions between endothelial cells of lymphatic vessels." J Exp Med **204**(10): 2349-62.

Baluk, P., T. Tammela, et al. (2005). "Pathogenesis of persistent lymphatic vessel hyperplasia in chronic airway inflammation." J Clin Invest **115**(2): 247-57.

Bamforth, S. D., U. Kniesel, et al. (1999). "A dominant mutant of occludin disrupts tight junction structure and function." J Cell Sci **112** (Pt 12): 1879-88.

Barleon, B., S. Sozzani, et al. (1996). "Migration of human monocytes in response to vascular endothelial growth factor (VEGF) is mediated via the VEGF receptor flt-1." Blood **87**(8): 3336-43.

Bates, D. O. and F. E. Curry (1996). "Vascular endothelial growth factor increases hydraulic conductivity of isolated perfused microvessels." Am J Physiol **271**(6 Pt 2): H2520-8.

Baudino, T. A., C. McKay, et al. (2002). "c-Myc is essential for vasculogenesis and angiogenesis during development and tumor progression." Genes Dev **16**(19): 2530-43.

Bazzoni, G. and E. Dejana (2004). "Endothelial cell-to-cell junctions: molecular organization and role in vascular homeostasis." Physiol Rev **84**(3): 869-901.

Beil, M., H. Braxmeier, et al. (2005). "Quantitative analysis of keratin filament networks in scanning electron microscopy images of cancer cells." J Microsc **220**(Pt 2): 84-95.

Bello-Fernandez, C., G. Packham, et al. (1993). "The ornithine decarboxylase gene is a transcriptional target of c-Myc." Proc Natl Acad Sci U S A **90**(16): 7804-8.

Benjamin, L. E., D. Golijanin, et al. (1999). "Selective ablation of immature blood vessels in established human tumors follows vascular endothelial growth factor withdrawal." J Clin Invest **103**(2): 159-65.

Bouchard, C., K. Thieke, et al. (1999). "Direct induction of cyclin D2 by Myc contributes to cell cycle progression and sequestration of p27." Embo J **18**(19): 5321-33.

Brandvold, K. A., P. Neiman, et al. (2000). "Angiogenesis is an early event in the generation of myc-induced lymphomas." Oncogene **19**(23): 2780-5.

Breit, S., K. Ashman, et al. (2000). "The N-myc oncogene in human neuroblastoma cells: down-regulation of an angiogenesis inhibitor identified as activin A." Cancer Res **60**(16): 4596-601.

Carmeliet, P. (2000). "Mechanisms of angiogenesis and arteriogenesis." Nat Med **6**(4): 389-95.

Carmeliet, P., V. Ferreira, et al. (1996). "Abnormal blood vessel development and lethality in embryos lacking a single VEGF allele." Nature **380**(6573): 435-9.

Caron, K. M. and O. Smithies (2001). "Extreme hydrops fetalis and cardiovascular abnormalities in mice lacking a functional Adrenomedullin gene." Proc Natl Acad Sci U S A **98**(2): 615-9.

Cera, M. R., A. Del Prete, et al. (2004). "Increased DC trafficking to lymph nodes and contact hypersensitivity in junctional adhesion molecule-A-deficient mice." J Clin Invest **114**(5): 729-38.

Clauss, M., H. Weich, et al. (1996). "The vascular endothelial growth factor receptor Flt-1 mediates biological activities. Implications for a functional role of placenta growth factor in monocyte activation and chemotaxis." J Biol Chem **271**(30): 17629-34.

Claxton, S. and M. Fruttiger (2004). "Periodic Delta-like 4 expression in developing retinal arteries." Gene Expr Patterns **5**(1): 123-7.

Coller, H. A., C. Grandori, et al. (2000). "Expression analysis with oligonucleotide microarrays reveals that MYC regulates genes involved in growth, cell cycle, signaling, and adhesion." Proc Natl Acad Sci U S A **97**(7): 3260-5.

Connolly, D. T., D. M. Heuvelman, et al. (1989). "Tumor vascular permeability factor stimulates endothelial cell growth and angiogenesis." J Clin Invest **84**(5): 1470-8.

Cooke, V. G., M. U. Naik, et al. (2006). "Fibroblast growth factor-2 failed to induce angiogenesis in junctional adhesion molecule-A-deficient mice." Arterioscler Thromb Vasc Biol **26**(9): 2005-11.

Costell, M., E. Gustafsson, et al. (1999). "Perlecan maintains the integrity of cartilage and some basement membranes." J Cell Biol **147**(5): 1109-22.

Crosby, J. R., R. A. Seifert, et al. (1998). "Chimaeric analysis reveals role of Pdgf receptors in all muscle lineages." Nat Genet **18**(4): 385-8.

Cursiefen, C., L. Chen, et al. (2004). "VEGF-A stimulates lymphangiogenesis and hemangiogenesis in inflammatory neovascularization via macrophage recruitment." J Clin Invest **113**(7): 1040-50.

Dackor, R. T., K. Fritz-Six, et al. (2006). "Hydrops fetalis, cardiovascular defects, and embryonic lethality in mice lacking the calcitonin receptor-like receptor gene." Mol Cell Biol **26**(7): 2511-8.

Dalla-Favera, R., E. P. Gelmann, et al. (1982). "Cloning and characterization of different human sequences related to the onc gene (v-myc) of avian myelocytomatosis virus (MC29)." Proc Natl Acad Sci U S A **79**(21): 6497-501.

Daly, C., E. Pasnikowski, et al. (2006). "Angiopoietin-2 functions as an autocrine protective factor in stressed endothelial cells." Proc Natl Acad Sci U S A **103**(42): 15491-6.

Davis, A. C., M. Wims, et al. (1993). "A null c-myc mutation causes lethality before 10.5 days of gestation in homozygotes and reduced fertility in heterozygous female mice." Genes Dev **7**(4): 671-82.

D'Cruz, C. M., E. J. Gunther, et al. (2001). "c-MYC induces mammary tumorigenesis by means of a preferred pathway involving spontaneous Kras2 mutations." Nat Med **7**(2): 235-9.

De Palma, M., M. A. Venneri, et al. (2005). "Tie2 identifies a hematopoietic lineage of proangiogenic monocytes required for tumor vessel formation and a mesenchymal population of pericyte progenitors." Cancer Cell **8**(3): 211-26.

De Palma, M., M. A. Venneri, et al. (2003). "Targeting exogenous genes to tumor angiogenesis by transplantation of genetically modified hematopoietic stem cells." Nat Med **9**(6): 789-95.

Dejana, E., F. Orsenigo, et al. (2008). "The role of adherens junctions and VE-cadherin in the control of vascular permeability." J Cell Sci **121**(Pt 13): 2115-22.

Deutsch, U., T. M. Schlaeger, et al. (2008). "Inducible endothelial cell-specific gene expression in transgenic mouse embryos and adult mice." Exp Cell Res **314**(6): 1202-16.

Dube, M., P. Madarnas, et al. (1992). "An animal model of Kaposi's sarcoma. I. Immune status of CD1 mice undergoing dimethyl hydrazine treatment to induce angiosarcomas and other malignancies." Anticancer Res **12**(1): 105-12.

Dubois, N. C., C. Adolphe, et al. (2008). "Placental rescue reveals a sole requirement for c-Myc in embryonic erythroblast survival and hematopoietic stem cell function." Development **135**(14): 2455-65.

Dumont, D. J., G. H. Fong, et al. (1995). "Vascularization of the mouse embryo: a study of flk-1, tek, tie, and vascular endothelial growth factor expression during development." Dev Dyn **203**(1): 80-92.

Dumont, D. J., G. Gradwohl, et al. (1994). "Dominant-negative and targeted null mutations in the endothelial receptor tyrosine kinase, tek, reveal a critical role in vasculogenesis of the embryo." Genes Dev **8**(16): 1897-909.

Dumont, D. J., L. Jussila, et al. (1998). "Cardiovascular failure in mouse embryos deficient in VEGF receptor-3." Science **282**(5390): 946-9.

Dunworth, W. P., K. L. Fritz-Six, et al. (2008). "Adrenomedullin stabilizes the lymphatic endothelial barrier in vitro and in vivo." Peptides **29**(12): 2243-9.

Egle, A., A. W. Harris, et al. (2004). "Bim is a suppressor of Myc-induced mouse B cell leukemia." Proc Natl Acad Sci U S A **101**(16): 6164-9.

Eischen, C. M., D. Woo, et al. (2001). "Apoptosis triggered by Myc-induced suppression of Bcl-X(L) or Bcl-2 is bypassed during lymphomagenesis." Mol Cell Biol **21**(15): 5063-70.

Eisenman, R. N. (2001). "Deconstructing myc." Genes Dev **15**(16): 2023-30.

Engelhardt, B. and H. Wolburg (2004). "Mini-review: Transendothelial migration of leukocytes: through the front door or around the side of the house?" Eur J Immunol **34**(11): 2955-63.

Esser, S., M. G. Lampugnani, et al. (1998). "Vascular endothelial growth factor induces VE-cadherin tyrosine phosphorylation in endothelial cells." J Cell Sci **111 (Pt 13)**: 1853-65.

Evan, G. and T. Littlewood (1998). "A matter of life and cell death." Science **281**(5381): 1317-22.

Evan, G. I. and K. H. Vousden (2001). "Proliferation, cell cycle and apoptosis in cancer." Nature **411**(6835): 342-8.

Evan, G. I., A. H. Wyllie, et al. (1992). "Induction of apoptosis in fibroblasts by c-myc protein." Cell **69**(1): 119-28.

Felsher, D. W. and J. M. Bishop (1999a). "Reversible tumorigenesis by MYC in hematopoietic lineages." Mol Cell **4**(2): 199-207.

Felsher, D. W. and J. M. Bishop (1999b). "Transient excess of MYC activity can elicit genomic instability and tumorigenesis." Proc Natl Acad Sci U S A **96**(7): 3940-4.

Fernandez-Sauze, S., C. Delfino, et al. (2004). "Effects of adrenomedullin on endothelial cells in the multistep process of angiogenesis: involvement of CRLR/RAMP2 and CRLR/RAMP3 receptors." Int J Cancer **108**(6): 797-804.

Ferrara, N., K. Carver-Moore, et al. (1996). "Heterozygous embryonic lethality induced by targeted inactivation of the VEGF gene." Nature **380**(6573): 439-42.

Fiedler, U. and H. G. Augustin (2006). "Angiopoietins: a link between angiogenesis and inflammation." Trends Immunol **27**(12): 552-8.

Foley, K. P. and R. N. Eisenman (1999). "Two MAD tails: what the recent knockouts of Mad1 and Mxi1 tell us about the MYC/MAX/MAD network." Biochim Biophys Acta **1423**(3): M37-47.

Fong, G. H., J. Rossant, et al. (1995). "Role of the Flt-1 receptor tyrosine kinase in regulating the assembly of vascular endothelium." Nature **376**(6535): 66-70.

Fong, G. H., L. Zhang, et al. (1999). "Increased hemangioblast commitment, not vascular disorganization, is the primary defect in flt-1 knock-out mice." Development **126**(13): 3015-25.

Fotsis, T., S. Breit, et al. (1999). "Down-regulation of endothelial cell growth inhibitors by enhanced MYCN oncogene expression in human neuroblastoma cells." Eur J Biochem **263**(3): 757-64.

Fox, J. G. (2007). The Mouse in Biomedical Research, American College of Laboratory.

Fritz-Six, K. L., W. P. Dunworth, et al. (2008). "Adrenomedullin signaling is necessary for murine lymphatic vascular development." J Clin Invest **118**(1): 40-50.

Fukuhara, S., K. Sako, et al. (2009). "Tie2 is tied at the cell-cell contacts and to extracellular matrix by angiopoietin-1." Exp Mol Med **41**(3): 133-9.

Gale, N. W., G. Thurston, et al. (2002). "Angiopoietin-2 is required for postnatal angiogenesis and lymphatic patterning, and only the latter role is rescued by Angiopoietin-1." Dev Cell **3**(3): 411-23.

Gallant, P., Y. Shiio, et al. (1996). "Myc and Max homologs in Drosophila." Science **274**(5292): 1523-7.

Gamble, J. R., J. Drew, et al. (2000). "Angiopoietin-1 is an antipermeability and anti-inflammatory agent in vitro and targets cell junctions." Circ Res **87**(7): 603-7.

Gardner, L., Lee, L., Dang, C (2002). The c-Myc Oncogenic Transcription Factor. The Encyclopedia of Cancer.

Gerber, H. P., K. J. Hillan, et al. (1999). "VEGF is required for growth and survival in neonatal mice." Development **126**(6): 1149-59.

Gerhardt, H. (2008). "VEGF and endothelial guidance in angiogenic sprouting." Organogenesis **4**(4): 241-6.

Gerhardt, H., M. Golding, et al. (2003). "VEGF guides angiogenic sprouting utilizing endothelial tip cell filopodia." J Cell Biol **161**(6): 1163-77.

Gossen, M., S. Freundlieb, et al. (1995). "Transcriptional activation by tetracyclines in mammalian cells." Science **268**(5218): 1766-9.

Grandori, C., S. M. Cowley, et al. (2000). "The Myc/Max/Mad network and the transcriptional control of cell behavior." Annu Rev Cell Dev Biol **16**: 653-99.

Gridley, T. (2007). "Notch signaling in vascular development and physiology." Development **134**(15): 2709-18.

Hackett, S. F., S. Wiegand, et al. (2002). "Angiopoietin-2 plays an important role in retinal angiogenesis." J Cell Physiol **192**(2): 182-7.

Handa, S., A. M. Sadi, et al. (2008). "Region-specific patterns of vascular remodelling occur early in atherosclerosis and without loss of smooth muscle cell markers." Atherosclerosis **196**(2): 617-23.

Hayashi, K., R. Makino, et al. (1987). "Characterization of rat c-myc and adjacent regions." Nucleic Acids Res **15**(16): 6419-36.

He, C., H. Hu, et al. (2008). "c-myc in the hematopoietic lineage is crucial for its angiogenic function in the mouse embryo." Development **135**(14): 2467-77.

Hellstrom, M., H. Gerhardt, et al. (2001). "Lack of pericytes leads to endothelial hyperplasia and abnormal vascular morphogenesis." J Cell Biol **153**(3): 543-53.

Hellstrom, M., M. Kalen, et al. (1999). "Role of PDGF-B and PDGFR-beta in recruitment of vascular smooth muscle cells and pericytes during embryonic blood vessel formation in the mouse." Development **126**(14): 3047-55.

Hellstrom, M., L. K. Phng, et al. (2007). "Dll4 signalling through Notch1 regulates formation of tip cells during angiogenesis." Nature **445**(7129): 776-80.

Hermeking, H. and D. Eick (1994). "Mediation of c-Myc-induced apoptosis by p53." Science **265**(5181): 2091-3.

Hermeking, H., C. Rago, et al. (2000). "Identification of CDK4 as a target of c-MYC." Proc Natl Acad Sci U S A **97**(5): 2229-34.

Herold, S., M. Wanzel, et al. (2002). "Negative regulation of the mammalian UV response by Myc through association with Miz-1." Mol Cell **10**(3): 509-21.

Hippenstiel, S., M. Witzenrath, et al. (2002). "Adrenomedullin reduces endothelial hyperpermeability." Circ Res **91**(7): 618-25.

Hirakawa, S., S. Kodama, et al. (2005). "VEGF-A induces tumor and sentinel lymph node lymphangiogenesis and promotes lymphatic metastasis." J Exp Med **201**(7): 1089-99.

Hiratsuka, S., O. Minowa, et al. (1998). "Flt-1 lacking the tyrosine kinase domain is sufficient for normal development and angiogenesis in mice." Proc Natl Acad Sci U S A **95**(16): 9349-54.

Hobson, B. and J. Denekamp (1984). "Endothelial proliferation in tumours and normal tissues: continuous labelling studies." Br J Cancer **49**(4): 405-13.

Hocke, A. C., B. Temmesfeld-Wollbrueck, et al. (2006). "Perturbation of endothelial junction proteins by Staphylococcus aureus alpha-toxin: inhibition of endothelial gap formation by adrenomedullin." Histochem Cell Biol **126**(3): 305-16.

Hofmann, J. J. and M. Luisa Iruela-Arispe (2007). "Notch expression patterns in the retina: An eye on receptor-ligand distribution during angiogenesis." Gene Expr Patterns **7**(4): 461-70.

Hong, Y. K., B. Lange-Asschenfeldt, et al. (2004). "VEGF-A promotes tissue repair-associated lymphatic vessel formation via VEGFR-2 and the alpha1beta1 and alpha2beta1 integrins." Faseb J **18**(10): 1111-3.

Ichikawa-Shindo, Y., T. Sakurai, et al. (2008). "The GPCR modulator protein RAMP2 is essential for angiogenesis and vascular integrity." J Clin Invest **118**(1): 29-39.

Ingvarsson, S., C. Asker, et al. (1988). "Structure and expression of B-myc, a new member of the myc gene family." Mol Cell Biol **8**(8): 3168-74.

Jain, M., C. Arvanitis, et al. (2002). "Sustained loss of a neoplastic phenotype by brief inactivation of MYC." Science **297**(5578): 102-4.

Janz, A., C. Sevignani, et al. (2000). "Activation of the myc oncoprotein leads to increased turnover of thrombospondin-1 mRNA." Nucleic Acids Res **28**(11): 2268-75.

Jeltsch, M., A. Kaipainen, et al. (1997). "Hyperplasia of lymphatic vessels in VEGF-C transgenic mice." Science **276**(5317): 1423-5.

Jeon, B. H., C. Jang, et al. (2008). "Profound but dysfunctional lymphangiogenesis via vascular endothelial growth factor ligands from CD11b+ macrophages in advanced ovarian cancer." Cancer Res **68**(4): 1100-9.

Juin, P., A. O. Hueber, et al. (1999). "c-Myc-induced sensitization to apoptosis is mediated through cytochrome c release." Genes Dev **13**(11): 1367-81.

Kahn, M. L. (2008). "Blood is thicker than lymph." J Clin Invest **118**(1): 23-6.

Kaipainen, A., J. Korhonen, et al. (1995). "Expression of the fms-like tyrosine kinase 4 gene becomes restricted to lymphatic endothelium during development." Proc Natl Acad Sci U S A **92**(8): 3566-70.

Kappel, A., V. Ronicke, et al. (1999). "Identification of vascular endothelial growth factor (VEGF) receptor-2 (Flk-1) promoter/enhancer sequences sufficient for angioblast and endothelial cell-specific transcription in transgenic mice." Blood **93**(12): 4284-92.

Karkkainen, M. J., P. Haiko, et al. (2004). "Vascular endothelial growth factor C is required for sprouting of the first lymphatic vessels from embryonic veins." Nat Immunol **5**(1): 74-80.

Karlsson, A., D. Deb-Basu, et al. (2003a). "Defective double-strand DNA break repair and chromosomal translocations by MYC overexpression." Proc Natl Acad Sci U S A **100**(17): 9974-9.

Karlsson, A., S. Giuriato, et al. (2003b). "Genomically complex lymphomas undergo sustained tumor regression upon MYC inactivation unless they acquire novel chromosomal translocations." Blood **101**(7): 2797-803.

Karpanen, T. and K. Alitalo (2008). "Molecular biology and pathology of lymphangiogenesis." Annu Rev Pathol **3**: 367-97.

Karpanen, T. and T. Makinen (2006). "Regulation of lymphangiogenesis--from cell fate determination to vessel remodeling." Exp Cell Res **312**(5): 575-83.

Kato, G. J. and C. V. Dang (1992). "Function of the c-Myc oncoprotein." Faseb J **6**(12): 3065-72.

Katsuno, T., K. Umeda, et al. (2008). "Deficiency of zonula occludens-1 causes embryonic lethal phenotype associated with defected yolk sac angiogenesis and apoptosis of embryonic cells." Mol Biol Cell **19**(6): 2465-75.

Kim, I., J. H. Kim, et al. (2000). "Angiopoietin-2 at high concentration can enhance endothelial cell survival through the phosphatidylinositol 3'-kinase/Akt signal transduction pathway." Oncogene **19**(39): 4549-52.

Korhonen, J., J. Partanen, et al. (1992). "Enhanced expression of the tie receptor tyrosine kinase in endothelial cells during neovascularization." Blood **80**(10): 2548-55.

Lemaitre, J. M., R. S. Buckle, et al. (1996). "c-Myc in the control of cell proliferation and embryonic development." Adv Cancer Res **70**: 95-144.

Lindahl, P., M. Hellstrom, et al. (1998). "Endothelial-perivascular cell signaling in vascular development: lessons from knockout mice." Curr Opin Lipidol **9**(5): 407-11.

Lindahl, P., B. R. Johansson, et al. (1997). "Pericyte loss and microaneurysm formation in PDGF-B-deficient mice." Science **277**(5323): 242-5.

Lobov, I. B., P. C. Brooks, et al. (2002). "Angiopoietin-2 displays VEGF-dependent modulation of capillary structure and endothelial cell survival in vivo." Proc Natl Acad Sci U S A **99**(17): 11205-10.

Lobov, I. B., R. A. Renard, et al. (2007). "Delta-like ligand 4 (Dll4) is induced by VEGF as a negative regulator of angiogenic sprouting." Proc Natl Acad Sci U S A **104**(9): 3219-24.

Lowe, S. W., E. Cepero, et al. (2004). "Intrinsic tumour suppression." Nature **432**(7015): 307-15.

Lum, H. and A. B. Malik (1994). "Regulation of vascular endothelial barrier function." Am J Physiol **267**(3 Pt 1): L223-41.

Madarnas, P., M. Dube, et al. (1992). "An animal model of Kaposi's sarcoma. II. Pathogenesis of dimethyl hydrazine induced angiosarcoma and colorectal cancer in three mouse strains." Anticancer Res **12**(1): 113-7.

Magid, R., T. J. Murphy, et al. (2003). "Expression of matrix metalloproteinase-9 in endothelial cells is differentially regulated by shear stress. Role of c-Myc." J Biol Chem **278**(35): 32994-9.

Maisonpierre, P. C., C. Suri, et al. (1997). "Angiopoietin-2, a natural antagonist for Tie2 that disrupts in vivo angiogenesis." Science **277**(5322): 55-60.

Makinen, T., L. Jussila, et al. (2001a). "Inhibition of lymphangiogenesis with resulting lymphedema in transgenic mice expressing soluble VEGF receptor-3." Nat Med **7**(2): 199-205.

Makinen, T., T. Veikkola, et al. (2001b). "Isolated lymphatic endothelial cells transduce growth, survival and migratory signals via the VEGF-C/D receptor VEGFR-3." Embo J **20**(17): 4762-73.

Marelli-Berg, F. M., E. Peek, et al. (2000). "Isolation of endothelial cells from murine tissue." J Immunol Methods **244**(1-2): 205-15.

Marinkovic, D., T. Marinkovic, et al. (2004). "Reversible lymphomagenesis in conditionally c-MYC expressing mice." Int J Cancer **110**(3): 336-42.

Martin, V., D. Liu, et al. (2008). "Tie2: a journey from normal angiogenesis to cancer and beyond." Histol Histopathol **23**(6): 773-80.

Mehta, D. and A. B. Malik (2006). "Signaling mechanisms regulating endothelial permeability." Physiol Rev **86**(1): 279-367.

Millan, J., L. Hewlett, et al. (2006a). "Lymphocyte transcellular migration occurs through recruitment of endothelial ICAM-1 to caveola- and F-actin-rich domains." Nat Cell Biol **8**(2): 113-23.

Millan, J., L. Williams, et al. (2006b). "An in vitro model to study the role of endothelial rho GTPases during leukocyte transendothelial migration." Methods Enzymol **406**: 643-55.

Miquerol, L., B. L. Langille, et al. (2000). "Embryonic development is disrupted by modest increases in vascular endothelial growth factor gene expression." Development **127**(18): 3941-6.

Mitchell, K. O., M. S. Ricci, et al. (2000). "Bax is a transcriptional target and mediator of c-myc-induced apoptosis." Cancer Res **60**(22): 6318-25.

Miyashita, K., H. Itoh, et al. (2003a). "Adrenomedullin promotes proliferation and migration of cultured endothelial cells." Hypertens Res **26 Suppl**: S93-8.

Miyashita, K., H. Itoh, et al. (2003b). "Adrenomedullin provokes endothelial Akt activation and promotes vascular regeneration both in vitro and in vivo." FEBS Lett **544**(1-3): 86-92.

Montaner, S., A. Sodhi, et al. (2003). "Endothelial infection with KSHV genes in vivo reveals that vGPCR initiates Kaposi's sarcomagenesis and can promote the tumorigenic potential of viral latent genes." Cancer Cell **3**(1): 23-36.

Nagy, J. A., E. Vasile, et al. (2002). "Vascular permeability factor/vascular endothelial growth factor induces lymphangiogenesis as well as angiogenesis." J Exp Med **196**(11): 1497-506.

Nau, M. M., B. J. Brooks, et al. (1985). "L-myc, a new myc-related gene amplified and expressed in human small cell lung cancer." Nature **318**(6041): 69-73.

Ngo, C. V., M. Gee, et al. (2000). "An in vivo function for the transforming Myc protein: elicitation of the angiogenic phenotype." Cell Growth Differ **11**(4): 201-10.

Nieminen, M., T. Henttinen, et al. (2006). "Vimentin function in lymphocyte adhesion and transcellular migration." Nat Cell Biol **8**(2): 156-62.

Nikiforov, M. A., S. Chandriani, et al. (2002). "A functional screen for Myc-responsive genes reveals serine hydroxymethyltransferase, a major source of the one-carbon unit for cell metabolism." Mol Cell Biol **22**(16): 5793-800.

Nilsson, J. A. and J. L. Cleveland (2003). "Myc pathways provoking cell suicide and cancer." Oncogene **22**(56): 9007-21.

Nitta, T., M. Hata, et al. (2003). "Size-selective loosening of the blood-brain barrier in claudin-5-deficient mice." J Cell Biol **161**(3): 653-60.

Noguera-Troise, I., C. Daly, et al. (2006). "Blockade of Dll4 inhibits tumour growth by promoting non-productive angiogenesis." Nature **444**(7122): 1032-7.

Nyqvist, D., C. Giampietro, et al. (2008). "Deciphering the functional role of endothelial junctions by using in vivo models." EMBO Rep **9**(8): 742-7.

Okajima, E. and U. P. Thorgeirsson (2000). "Different regulation of vascular endothelial growth factor expression by the ERK and p38 kinase pathways in v-ras, v-raf, and v-myc transformed cells." Biochem Biophys Res Commun **270**(1): 108-11.

Oliver, G. (2004). "Lymphatic vasculature development." Nat Rev Immunol **4**(1): 35-45.

Oliver, G. and K. Alitalo (2005). "The lymphatic vasculature: recent progress and paradigms." Annu Rev Cell Dev Biol **21**: 457-83.

Oshima, Y., T. Deering, et al. (2004). "Angiopoietin-2 enhances retinal vessel sensitivity to vascular endothelial growth factor." J Cell Physiol **199**(3): 412-7.

Oshima, Y., S. Oshima, et al. (2005). "Different effects of angiopoietin-2 in different vascular beds: new vessels are most sensitive." Faseb J **19**(8): 963-5.

Partanen, J. and D. J. Dumont (1999). "Functions of Tie1 and Tie2 receptor tyrosine kinases in vascular development." Curr Top Microbiol Immunol **237**: 159-72.

Partridge, C. A., C. J. Horvath, et al. (1992). "Influence of extracellular matrix in tumor necrosis factor-induced increase in endothelial permeability." Am J Physiol **263**(6 Pt 1): L627-33.

Passi, A., D. Negrini, et al. (1999). "The sensitivity of versican from rabbit lung to gelatinase A (MMP-2) and B (MMP-9) and its involvement in the development of hydraulic lung edema." FEBS Lett **456**(1): 93-6.

Patel, J. H., A. P. Loboda, et al. (2004). "Analysis of genomic targets reveals complex functions of MYC." Nat Rev Cancer **4**(7): 562-8.

Pelengaris, S., M. Khan, et al. (2002a). "c-MYC: more than just a matter of life and death." Nat Rev Cancer **2**(10): 764-76.

Pelengaris, S., M. Khan, et al. (2002b). "Suppression of Myc-induced apoptosis in beta cells exposes multiple oncogenic properties of Myc and triggers carcinogenic progression." Cell **109**(3): 321-34.

Pelengaris, S., T. Littlewood, et al. (1999). "Reversible activation of c-Myc in skin: induction of a complex neoplastic phenotype by a single oncogenic lesion." Mol Cell **3**(5): 565-77.

Peters, K. G., A. Coogan, et al. (1998). "Expression of Tie2/Tek in breast tumour vasculature provides a new marker for evaluation of tumour angiogenesis." Br J Cancer **77**(1): 51-6.

Peters, K. G., C. D. Kontos, et al. (2004). "Functional significance of Tie2 signaling in the adult vasculature." Recent Prog Horm Res **59**: 51-71.

Petrova, T. V., T. Makinen, et al. (1999). "Signaling via vascular endothelial growth factor receptors." Exp Cell Res **253**(1): 117-30.

Poole, T. J., E. B. Finkelstein, et al. (2001). "The role of FGF and VEGF in angioblast induction and migration during vascular development." Dev Dyn **220**(1): 1-17.

Puri, M. C., J. Partanen, et al. (1999). "Interaction of the TEK and TIE receptor tyrosine kinases during cardiovascular development." Development **126**(20): 4569-80.

Ridgway, J., G. Zhang, et al. (2006). "Inhibition of Dll4 signalling inhibits tumour growth by deregulating angiogenesis." Nature **444**(7122): 1083-7.

Ridler, C. S. (1978). "Thresholding Using an Iterative Selection Method." IEEE Trans. System, Man and Cybernetics **8**: 630-632.

Risau, W. (1997). "Mechanisms of angiogenesis." Nature **386**(6626): 671-4.

Risau, W. and I. Flamme (1995). "Vasculogenesis." Annu Rev Cell Dev Biol **11**: 73-91.

Roberts, W. G. and G. E. Palade (1995). "Increased microvascular permeability and endothelial fenestration induced by vascular endothelial growth factor." J Cell Sci **108 (Pt 6)**: 2369-79.

Rossant, J. and L. Howard (2002). "Signaling pathways in vascular development." Annu Rev Cell Dev Biol **18**: 541-73.

Rundhaug, J. E. (2005). "Matrix metalloproteinases and angiogenesis." J Cell Mol Med **9**(2): 267-85.

Saitou, M., M. Furuse, et al. (2000). "Complex phenotype of mice lacking occludin, a component of tight junction strands." Mol Biol Cell **11**(12): 4131-42.

Sato, T. N., Y. Tozawa, et al. (1995). "Distinct roles of the receptor tyrosine kinases Tie-1 and Tie-2 in blood vessel formation." Nature **376**(6535): 70-4.

Sawano, A., S. Iwai, et al. (2001). "Flt-1, vascular endothelial growth factor receptor 1, is a novel cell surface marker for the lineage of monocyte-macrophages in humans." Blood **97**(3): 785-91.

Scehnet, J. S., W. Jiang, et al. (2007). "Inhibition of Dll4-mediated signaling induces proliferation of immature vessels and results in poor tissue perfusion." Blood **109**(11): 4753-60.

Scharpfenecker, M., U. Fiedler, et al. (2005). "The Tie-2 ligand angiopoietin-2 destabilizes quiescent endothelium through an internal autocrine loop mechanism." J Cell Sci **118**(Pt 4): 771-80.

Schmidt, T. G., Lang F (2000). Physiologie des Menschen, Springer.

Schwab, M., K. Alitalo, et al. (1983). "Amplified DNA with limited homology to myc cellular oncogene is shared by human neuroblastoma cell lines and a neuroblastoma tumour." Nature **305**(5931): 245-8.

Seetharam, L., N. Gotoh, et al. (1995). "A unique signal transduction from FLT tyrosine kinase, a receptor for vascular endothelial growth factor VEGF." Oncogene **10**(1): 135-47.

Senger, D. R., S. J. Galli, et al. (1983). "Tumor cells secrete a vascular permeability factor that promotes accumulation of ascites fluid." Science **219**(4587): 983-5.

Seoane, J., H. V. Le, et al. (2002). "Myc suppression of the p21(Cip1) Cdk inhibitor influences the outcome of the p53 response to DNA damage." Nature **419**(6908): 729-34.

Shachaf, C. M., A. M. Kopelman, et al. (2004). "MYC inactivation uncovers pluripotent differentiation and tumour dormancy in hepatocellular cancer." Nature **431**(7012): 1112-7.

Shalaby, F., J. Rossant, et al. (1995). "Failure of blood-island formation and vasculogenesis in Flk-1-deficient mice." Nature **376**(6535): 62-6.

Sherr, C. J. and J. D. Weber (2000). "The ARF/p53 pathway." Curr Opin Genet Dev **10**(1): 94-9.

Shibuya, M. and L. Claesson-Welsh (2006). "Signal transduction by VEGF receptors in regulation of angiogenesis and lymphangiogenesis." Exp Cell Res **312**(5): 549-60.

Shim, H., C. Dolde, et al. (1997). "c-Myc transactivation of LDH-A: implications for tumor metabolism and growth." Proc Natl Acad Sci U S A **94**(13): 6658-63.

Soille, P. (1999). Morphological Image Analysis. Berlin, Springer.

Solomon, B. L., Martin DW (2005). Biology, Brooks\Cole Thomson Learning.

Soucie, E. L., M. G. Annis, et al. (2001). "Myc potentiates apoptosis by stimulating Bax activity at the mitochondria." Mol Cell Biol **21**(14): 4725-36.

Stanton, L. W., P. D. Fahrlander, et al. (1984). "Nucleotide sequence comparison of normal and translocated murine c-myc genes." Nature **310**(5976): 423-5.

Suchting, S., C. Freitas, et al. (2007). "The Notch ligand Delta-like 4 negatively regulates endothelial tip cell formation and vessel branching." Proc Natl Acad Sci U S A **104**(9): 3225-30.

Sugiyama, A., A. Kume, et al. (1989). "Isolation and characterization of s-myc, a member of the rat myc gene family." Proc Natl Acad Sci U S A **86**(23): 9144-8.

Suri, C., P. F. Jones, et al. (1996). "Requisite role of angiopoietin-1, a ligand for the TIE2 receptor, during embryonic angiogenesis." Cell **87**(7): 1171-80.

Takahama, M., M. Tsutsumi, et al. (1999). "Enhanced expression of Tie2, its ligand angiopoietin-1, vascular endothelial growth factor, and CD31 in human non-small cell lung carcinomas." Clin Cancer Res **5**(9): 2506-10.

Takahashi, H. and M. Shibuya (2005). "The vascular endothelial growth factor (VEGF)/VEGF receptor system and its role under physiological and pathological conditions." Clin Sci (Lond) **109**(3): 227-41.

Tammela, T., A. Saaristo, et al. (2005). "Angiopoietin-1 promotes lymphatic sprouting and hyperplasia." Blood **105**(12): 4642-8.

Taylor AE, G. D. (1984). Exchange of macromolecules across the microcirculation. Handbook of physiology. Bethesda, MD, American Physiological Society. **Vol 4.**

Teichert-Kuliszewska, K., P. C. Maisonpierre, et al. (2001). "Biological action of angiopoietin-2 in a fibrin matrix model of angiogenesis is associated with activation of Tie2." Cardiovasc Res **49**(3): 659-70.

Thomas, M. and H. G. Augustin (2009). "The role of the Angiopoietins in vascular morphogenesis." Angiogenesis **12**(2): 125-37.

Thurston, G., J. S. Rudge, et al. (2000). "Angiopoietin-1 protects the adult vasculature against plasma leakage." Nat Med **6**(4): 460-3.

Thurston, G., C. Suri, et al. (1999). "Leakage-resistant blood vessels in mice transgenically overexpressing angiopoietin-1." Science **286**(5449): 2511-4.

Vafa, O., M. Wade, et al. (2002). "c-Myc can induce DNA damage, increase reactive oxygen species, and mitigate p53 function: a mechanism for oncogene-induced genetic instability." Mol Cell **9**(5): 1031-44.

Vandenbroucke, E., D. Mehta, et al. (2008). "Regulation of endothelial junctional permeability." Ann N Y Acad Sci **1123**: 134-45.

Vennstrom, B., D. Sheiness, et al. (1982). "Isolation and characterization of c-myc, a cellular homolog of the oncogene (v-myc) of avian myelocytomatosis virus strain 29." J Virol **42**(3): 773-9.

Vestweber, D. (2007). "Adhesion and signaling molecules controlling the transmigration of leukocytes through endothelium." Immunol Rev **218**: 178-96.

Volpert, O. V. and R. M. Alani (2003). "Wiring the angiogenic switch: Ras, Myc, and Thrombospondin-1." Cancer Cell **3**(3): 199-200.

Wagner, A. J., J. M. Kokontis, et al. (1994). "Myc-mediated apoptosis requires wild-type p53 in a manner independent of cell cycle arrest and the ability of p53 to induce p21waf1/cip1." Genes Dev **8**(23): 2817-30.

Weis, S. M. (2008). "Vascular permeability in cardiovascular disease and cancer." Curr Opin Hematol **15**(3): 243-9.

Witte, M. H., K. Jones, et al. (2006). "Structure function relationships in the lymphatic system and implications for cancer biology." Cancer Metastasis Rev **25**(2): 159-84.

Xu, J., P. J. Kausalya, et al. (2008). "Early embryonic lethality of mice lacking ZO-2, but Not ZO-3, reveals critical and nonredundant roles for individual zonula occludens proteins in mammalian development." Mol Cell Biol **28**(5): 1669-78.

Ziegler, B. L., M. Valtieri, et al. (1999). "KDR receptor: a key marker defining hematopoietic stem cells." Science **285**(5433): 1553-8.

Zindy, F., C. M. Eischen, et al. (1998). "Myc signaling via the ARF tumor suppressor regulates p53-dependent apoptosis and immortalization." Genes Dev **12**(15): 2424-33.

I want morebooks!

Buy your books fast and straightforward online - at one of world's fastest growing online book stores! Environmentally sound due to Print-on-Demand technologies.

Buy your books online at
www.morebooks.shop

Kaufen Sie Ihre Bücher schnell und unkompliziert online – auf einer der am schnellsten wachsenden Buchhandelsplattformen weltweit! Dank Print-On-Demand umwelt- und ressourcenschonend produziert.

Bücher schneller online kaufen
www.morebooks.shop

KS OmniScriptum Publishing
Brivibas gatve 197
LV-1039 Riga, Latvia
Telefax: +371 686 204 55

info@omniscriptum.com
www.omniscriptum.com

Printed by Books on Demand GmbH, Norderstedt / Germany